76 Advances in Polymer Science

Pharmacy/ Thermomechanics/ Elastomers/ Telechelics

Editor: K. Dušek

With Contributions by
J. B. Donnet, S. G. Entelis, V. V. Evreinov, G. Franz,
Y. K. Godovsky, A. V. Gorshkov, A. Vidal

With 84 Figures and 26 Tables

Springer-Verlag
Berlin Heidelberg New York Tokyo

ISBN-3-540-15830-8 Springer-Verlag Berlin Heidelberg New York Tokyo
ISBN-0-387-15830-8 Springer-Verlag New York Heidelberg Berlin Tokyo

Library of Congress Catalog Card Number 61-642

This work is subject to copyright. All rights are reserved, whether the whole or part of the material is concerned, specifically those of translation, reprinting, re-use of illustration, broadcasting, reproduction by photocopying machine or similar means, and storage in data banks. Under § 54 of the German Copyright Law where copies are made for other than private use, a fee is payable to "Verwertungsgesellschaft Wort", Munich.

© by Springer-Verlag Berlin · Heidelberg 1986
Printed in GDR

The use of registered names, trademarks, etc. in this publication does not imply, even in the absence of a specific statement, that such names are exempt from the relevant protective laws and regulations and therefore free for general use.

Typesetting and Offsetprinting: Th. Müntzer, GDR;
Bookbinding: Lüderitz & Bauer, Berlin
2154/3020-543210

Editors

Prof. Henri Benoit, CNRS, Centre de Recherches sur les Macromolecules, 6, rue Boussingault, 67083 Strasbourg Cedex, France

Prof. Hans-Joachim Cantow, Institut für Makromolekulare Chemie der Universität, Stefan-Meier-Str. 31, 7800 Freiburg i. Br., FRG

Prof. Gino Dall'Asta, Via Pusiano 30, 20137 Milano, Italy

Prof. Karel Dušek, Institute of Macromolecular Chemistry, Czechoslovak Academy of Sciences, 16206 Prague 616, ČSSR

Prof. John D. Ferry, Department of Chemistry, The University of Wisconsin, Madison, Wisconsin 53706, U.S.A.

Prof. Hiroshi Fujita, Department of Macromolecular Science, Osaka University, Toyonaka, Osaka, Japan

Prof. Manfred Gordon, Department of Pure Mathematics and Mathematical Statistics, University of Cambridge CB2 1SB, England

Prof. Gisela Henrici-Olivé, Chemical Department, University of California, San Diego, La Jolla, CA 92037, U.S.A.

Prof. Dr. habil. Günter Heublein, Sektion Chemie, Friedrich-Schiller-Universität, Humboldtstraße 10, 69 Jena, DDR

Prof. Dr. Hartwig Höcker, Universität Bayreuth, Makromolekulare Chemie I. Universitätsstr. 30, 8580 Bayreuth, FRG

Prof. Hans-Henning Kausch, Laboratoire de Polymères, École Polytechnique Fédérale de Lausanne, 32, ch. de Bellerive, 1007 Lausanne, Switzerland

Prof. Joseph P. Kennedy, Institute of Polymer Science, The University of Akron, Akron, Ohio 44325, U.S.A.

Prof. Anthony Ledwith, Department of Inorganic, Physical and Industrial Chemistry, University of Liverpool, Liverpool L69 3BX, England

Prof. Seizo Okamura, No. 24, Minamigoshi-Machi Okazaki, Sakyo-Ku Kyoto 606, Japan

Professor Salvador Olivé, Chemical Department, University of California, San Diego, La Jolla, CA 92037, U.S.A.

Prof. Charles G. Overberger, Department of Chemistry. The University of Michigan, Ann Arbor, Michigan 48 104, U.S.A.

Prof. Helmut Ringsdorf, Institut für Organische Chemie, Johannes-Gutenberg-Universität, J.-J.-Becher Weg 18–20, 6500 Mainz, FRG

Prof. Takeo Saegusa, Department of Synthetic Chemistry, Faculty of Engineering, Kyoto University, Kyoto, Japan

Prof. Günter Victor Schulz, Institut für Physikalische Chemie der Universität, 6500 Mainz, FRG

Prof. William P. Slichter, Chemical Physics Research Department, Bell Telephone Laboratories, Murray Hill, New Jersey 07971, U.S.A.

Prof. John K. Stille, Department of Chemistry. Colorado State University, Fort Collins, Colorado 80523, U.S.A.

Editorial

With the publication of Vol. 51 the editors and the publisher would like to take this opportunity to thank authors and readers for their collaboration and their efforts to meet the scientific requirements of this series. We appreciate the concern of our authors for the progress of "Advances in Polymer Science" and we also welcome the advice and critical comments of our readers.

With the publication of Vol. 51 we would also like to refer to a editorial policy: *this series publishes invited, critical review articles of new developments in all areas of polymer science in English (authors may naturally also include workes of their own)*. The responsible editor, that means the editor who has invited the author, discusses the scope of the review with the author on the basis of a tentative outline which the author is asked to provide. The author and editor are responsible for the scientific quality of the contribution.

Manuscripts must be submitted in content, language, and form satisfactory to Springer-Verlag. Figures and formulas should be reproducible. To meet the convenience of our readers, the publisher will include "volume index" which characterizes the content of the volume.

The editors and the publisher will make all efforts to publish the manuscripts as rapidly as possible, i.e., at the maximum six months after the submission of an accepted paper. Contributions from diverse areas of polymer science must occasionally be united in one volume. In such cases a "volume index" cannot meet all expectations, but will nevertheless provide more information than a mere volume number.

Starting with Vol. 51, each volume will contain a subject index.

Editors											Publisher

Table of Contents

Polysaccharides in Pharmacy
G. Franz 1

Thermomechanics of Polymers
Y. K. Godovsky 31

Carbon Black: Surface Properties and Interactions with Elastomers
J. B. Donnet, A. Vidal 103

Functionality and Molecular Weight Distribution of Telechelic Polymers
S. G. Entelis, V. V. Evreinov, A. V. Gorshkov 129

Author Index Volumes 1–76 177

Subject Index 187

Polysaccharides in Pharmacy

G. Franz
Department of Chemistry and Pharmacy, University of Regensburg,
8400 Regensburg/FRG

The present status of knowledge concerning the different applications of polysaccharides in the broad field of pharmacy is discussed. Starting with some general properties of polysaccharides, which are the base for industial use, the different classical pharmaceutical systems are presented and future dosage forms, which are currently in development, are discussed.

The most frequently employed polysaccharides and their derivatives are listed in the second part, Indications about biological origin, chemical structure, and the physicochemical properties, which are the basis for pharmaceutical use, are given.

In a last chapter, the usefulness of these biopolymers as substances influencing the immune system and acting on tumor cells is reviewed.

1 Introduction 3

2 General Properties of Polysaccharides 5

3 Polysaccharides in Different Pharmaceutical Systems 5
   3.1 Topical Systems 5
   3.2 Oral Dosage Forms: Solutions, Suspensions and Emulsions 6
   3.3 Solid Oral Dosage Forms 6
   3.4 Polysaccharides as Drug Carriers 8
   3.5 Ophthalmic Products 9
   3.6 Special Dosage Forms: Cyclodextrin Inclusion Complexes 10

4 Polysaccharides and their Respective Derivatives from Higher Plants . . . 11
   4.1 Cellulose and Derivatives 11
      4.1.1 Purified Cotton 11
      4.1.2 Cellulose Powder 11
      4.1.3 Methylcellulose 12
      4.1.4 Hydroxyethylcellulose 13
      4.1.5 Hydroxypropylmethylcellulose 13
      4.1.6 Oxidized Cellulose 13
      4.1.7 Cellulose Acetate Phthalate 14
      4.1.8 Carboxymethylcellulose (CMC) 14
   4.2 Starch and Derivatives 14
      4.2.1 Properties of Starch 14
      4.2.2 Composition of Starch 15
      4.2.3 Dextrin 16

    4.3 Gums and Mucilages 16
        4.3.1 Acacia Gum (Gum Arabic, Acaciate Gummi) 18
        4.3.2 Tragacanth Gum 19
        4.3.3 Sterculia Gum (Karaya Gum, Indian Tragacanth) 20
        4.3.4 Plantago Seed (Psyllium-Flea Seed) 20
        4.3.5 Quince Seed 20
    4.4 Pectic Substances 20
    4.5 Galactomannans 21
        4.5.1 Galactomannan Sources 22
    4.6 Xyloglucans 23

5 Polysaccharides from Algae 23
    5.1 Alginic Acid 23
    5.2 Agar 24
    5.3 Carrageenan 24
    5.4 Furcelleran 25

6 Polysaccharides from Bacteria 26
    6.1 Xanthan 26
    6.2 Dextran 26

7 Polysaccharides as Antitumor Agents 27
    7.1 Polysaccharide Structure and Antitumor Activity 28

8 References 29

1 Introduction

The use of various polysaccharides for pharmaceutical purposes has a long historical background and increased considerably during the last decades. This is not only the case for polysaccharides used as excipients, which are essential for specific formulations and constitute the most widely used form of polysaccharides, but also for a series of carbohydrate polymers which have proven to be most useful as direct drugs.

More recently experiments in the field of immunology and cancer research have shown that several polysaccharides are very promising nontoxic products. This has opened a completely new horizon for their applications as direct physiologically active substances. Compared to other applications in industry, the percentage of polysaccharides actually used for pharmaceutical purposes is rather low (Table 1).

When polysaccharides are used for internal applications, it has to be demonstrated that these substances are nontoxic. Teratogenic, mutagenic, allergenic, and other possible side effects have to be excluded (Table 2).

Since about the beginning of this century, compendial standards and government regulations require that all drug products, including carbohydrates, meet strict standards of identity, potency, and purity. These standards were thought to adequately define drug quality and have been enforced by legislation. Natural products may undergo changes with time resulting in a loss of biological and therapeutic activity. We now know that even polysaccharides require very careful evaluation to accurately and fully reflect their quality and performance. The actual performance of many drugs in clinical practice is known to be greatly affected by the method of delivery of the drug to the patient.

Factors affecting the delivery include: the physical form of the drug, the entry into the body, the design and the formula of the product. This may depend on the physicochemical properties of the excipients, the control of the drug-excipient interaction at the absorption site. Thus, polysaccharides may play an essential role for effectiveness and reliability of the different drug delivery systems.

With the exception of some linear polysaccharides such as cellulose, polysaccharides are hydrophilic polymers which, in a suitable solvent system, form hydrocolloids with different physical properties. Many polysaccharides are used mainly because of their

Table 1. Industrial usage of polysaccharides (Sandford and Baired [1])

Product type	Percentage
Laundry products	16
Textiles	14
Adhesives	12
Paper	10
Paint	9
Food	8
Pharmaceuticals	7
Other	24

Table 2. Important criteria for the pharmaceutical use of polysaccharides

Polysaccharides must have:	Polysaccharides must not:
— high purity	— destroy enzymes
— chemical, physical, and mechanical properties for proposed functions	— cause immuneresponses
	— cause cancer
— high stability	— produce toxic and allergic reactions
— easy fabricability	— deplete electrolytes

thickening and gelling properties. The presence of small amounts of these materials can drastically alter the rheological properties of solutions, thereby producing a change in the texture of the product. The differences in the mode of gelation, the quality and the stability of the gels, and the overall organoleptic properties of jellies differ widely.

The polysaccharides actually used for pharmaceutical purposes can be devided in two groups: the natural and the modified natural hydrocolloids (Table 3).

The large number of polysaccharides, which in recent years have been shown to be physiologically active as immunostimulants or antitumor drugs, are not included in Table 3. The origin and chemical nature of these compounds will be discussed below.

Table 3. Classification of hydrocolloids

Natural products	Modified natural products
Plant	*Cellulose derivatives*
— Arabic gum	— Carboxymethylcellulose
— Tragacanth gum	— Methylcellulose
— Karaya gum	— Hydroxyethylcellulose
— Ghatti gum	— Cellulose acetate phthalate
Plant seed polysaccharides	*Other derivatives*
— Guar	— modified starch
— Locust bean	— low-methyl pectin
— Psyllium	— propylene glycol alginate
— Tamarmid	
— Quince	
Plant extracts	
— Pectin	
— Arabinogalactan	
Cereal	
— Starches	
Sea weed extracts	
— Agar	
— Alginates	
— Carrageenans	
— Furcelleran	
Fermentation products	
— Dextran	
— Curdlan	
— Xanthan	

2 General Properties of Polysaccharides

Solubility is a very important criterium for the different uses of polysaccharides in pharmacy. Table 4 lists the solubility criteria of some polysaccharides in water. Solutions containing sugars and alcohol generally depress the solubility of polysaccharides. Polysaccharides containing carboxyl groups, i.e., pectins, alginates, and carboxymethylcellulose, are insoluble at low pH values. They will be precipitated when the pH is lowered below 3.

Pectins and alginates are generally insoluble in solvents containing divalent ions. An exception is the solubility of alginates in solutions containing magnesium ions. Solubilization of a polysaccharide can sometimes be achieved if a chelating agent is included in the formula. In general, the maximal concentration of a soluble polysaccharide is limited by the viscosity. Obviously, higher concentrations can be obtained with low-molecular-weight fractions. Polysaccharides in solution will depolymerize when the solution is heated and this will cause a drop in viscosity. The intensity of this phenomenon for any given time/temperature régime depends primarily on the polysaccharide structure and on the pH. Almost all polysaccharides depolymerize in solution, more rapidly in acid than in neutral media. An exception is xanthan gum which is extremely stable under all conditions, and pectin which is more stable in acid than in neutral media. Although depolymerization always causes a significant drop in viscosity, the gelling capacity may have only a small effect on gel rigidity.

Table 4. Solubility of polysaccharides in water (Pedersen [2])

	20 °C	100 °C
Guar gum	+	+
Locust gum	−	+
Pectin	+	+
Sodium alginate	+	+
Agar	−	+
Carrageenan	±	+
CMC	+	+
Microcrystalline cellulose	swelling	swelling
Xanthan	+	+

3 Polysaccharides in Different Pharmaceutical Systems

3.1 Topical Systems

Semisolid systems fulfill a special topical need by being able to cling to the surface of application. Such systems are plastic in behavior, which allows semisolids to be mechanically spread uniformly over a surface as an immobile film. For the production of lipid-free ointments, pastes, and creams, several gel-forming polysaccharides are being used. As an emulsifier they can provide a three-dimensional matrix which determines the extent and the quality of the gel-like micellar structure. The polymer is usually present in low concentrations ranging from 0.5–2%. Some systems are clear, others are turbid when the polymer is not fully dissolved or forming aggregates. This is the case with agar gels which are fluid at elevated temperatures, but solidify near room temperature due to an increase in chain interactions.

The natural polymers frequently used for the preparation of pharmaceutical gels include tragacanth, pectin, carrageenan, agar, and alginic acid, as well as semi-synthetic polysaccharides such as methylcellulose, hydroxymethylcellulose, and carboxymethylcellulose.

3.2 Oral Dosage Forms: Solutions, Suspensions, and Emulsions

Rheological properties of pharmaceutical disperse systems can be of particular importance for the specific use for which the product has been designed. Suspensions allow the development of a liquid dosage form containing an appropriate quantity of drug in a reasonably small volume. Suspensions can also mask the unpleasant taste of drugs. For the preparation of such a formula, suspending agents such as carboxymethylcellulose are often employed. However, care should be taken that there be no chemical or physicochemical interactions among suspending agent, surfactants, and the drug.

Besides suspensions, emulsions are dosage forms which have considerable traditional use in pharmacy. The technology of emulsions [oil-in-water (o/w) or water-in-oil (w/o)] is clearly established. In recent years, various emulsions such as water-oil in water have been developed, where water droplets are dispersed in oil droplets which are in turn dispersed in water. Such multiple emulsions may be used for prolonged oral action or intramuscular depot therapy.

Because of their large interfacial area, emulsions are basically unstable. In order to produce a stable emulsion, a surfactant is mostly needed. The surfactants are adsorbed at the oil-water interface, forming a link between the two phases of different polarity. For this purpose, a wide variety of emulsifying agents is currently available. Polysaccharides such as arabic gum, tragacanth, Karaya gum, and different seaweed carbohydrate polymers have been employed. They, however, show considerable batch-to-batch variations and might support microbial growth.

3.3 Solid Oral Dosage Forms

This group includes tablets, capsules, cachets, and pills as well as bulk or unit-dose powders and granules. The advantages of these solid forms are the ease of accurate dosage, good physical and chemical stability, and good appearance resulting in a high level of patient acceptability. Disadvantages might result from possible bioavailability problems caused by the fact that disintegration and dissolution must be completed before the drug is available for absorption. In order to overcome these problems, several polysaccharides have been employed to provide a controlled release of the drug or to allow a better rate of availability.

The most common solid dosage forms of granulated or powdered materials are prepared by compression. Many tablets are coated to minimize the unpleasant taste of certain substances or to protect the ingredients against decomposition and enhance the appearance.

In many cases, solid formulas are composed of one or more medicaments plus excipients of various types. Each component must be uniformly dispersed within the mixture and any tendency for component segregation has to be minimized. Furthermore, an increasing number of drugs are used in very low dosage, and to produce tablets of reasonable size, it is necessary to dilute the drug with an inert material such as a polysaccharide. It may be possible to combine the role of a diluent with a disintegrating agent. Starch, for example, is often used as a filler and as a disintegrating or binding agent. Disintegration is mostly due to capillary forces in producing a rapid uptake of aqueous liquids. As a result, the tablets swell in contact with water. Microcrystalline celluloses have been shown to be highly porous and, therefore, good disintegrators. They also serve as excellent binders by improving the mechanical strength of some weak formulations. Various cellulose derivatives have also been utilized for this purpose. Powdered gums such as Karaya, tragacanth, and agar can swell considerably, but their adhesiveness limits their value as disintegrators and restricts the concentrations to a maximum of about 5% of the tablet weight. Sodium alginate shows one of the best combinations of sufficient swelling with minimum adhesiveness. The relative degree of swelling of some polysaccharide disintegrator decreases in the following series: amylopectin, sodium alginate, carboxymethylcellulose, soluble starch, starch. In most cases, a granulation is required before a drug mixture with diluents, disintegrators, and additional granulating agents can be tabletted. Moist granulation, dry granulation, and preliminary compression are the commonly employed processes with suitable materials.

Often, a binding or adhesive granulating agent is required which may be added in solution or mixed with the tablet ingredients as a dry powder for subsequent activation by moisturing with an appropriate solvent. Many binders are hydrophilic colloids as, for example, polysaccharides (Table 5).

In order to reach a rapid disintegration after addition of a polysaccharide binder, addition of small quantities of appropriate enzymes may be useful (Table 6).

Table 5. Polysaccharides as adhesive granulating agents

Material	Concentration, % in dried granules
Arabic gum	2 –5
Galactomannan	1 –3
Sodium alginate	0.5–3
Tragacanth	0.5
Starch	2 –5
Methylcellulose	0.5–3
Ethylcellulose	0.5–3
Sodium carboxymethylcellulose	0.5–3

Table 6. Polysaccharide binders and enzymatic agents for disintegration

Binder	Enzymes
Starch	Amylose
Cellulose (derivatives)	Cellulase
Gums (Arabic gum, Tragacanth)	Hemicellulase
Alginates	Carrageenase

Special solid oral dosage forms are the controlled-release tablets. In some cases, it is desirable that the release of the drug from the tablet should be more gradual than in the normal case. This controlled slow release can be achieved in different ways.

In some formulations the drug is incorporated into an inert matrix, which only permits a release by diffusion mechanisms. For this purpose, some plant gums are used which, in a suitable medium, swell and produce a mucilaginous barrier to diffusion. Galactomannans from different biological sources have been tested and shown to be a very convenient neutral polymer (Nürnberg and Reij [3], Nürnberg and Bleimüller [4]).

3.4 Polysaccharides as Drug Carriers

Most medications are micromolecular in size and as such are fairly free to diffuse throughout the biological system. Consequently, drugs have been inherently difficult to administer in a localized mode in the target tissues or organs. Since polymers are often adsorbed at interfaces, the attachment of drugs to the backbone of a physiologically inactive macromolecule has been found to produce a polymer with a distinct pharmacological activity, such as sustained therapy, slow drug release, prolonged activity, and decreased drug metabolism and excretion. Besides a great number of synthetic polymers such as polyethylene, polystyrene, polyamides, and polysilicones, recently polysaccharides have been used for this purpose. Initially, cellulose and cellulose derivatives were activated and linked with active substrates (Malley and Campbell [5], Lilly et al. [6], Haimowich et al. [7]). Later on, dextrans of different molecular weights were used for similar purposes (Molteni and Scrolini [8], Marshal and Rabinowitz [9]). The ability of dextrans to form a variety of complexes with drugs is closely related to its basic chemical characteristics, such as high stability of the glycosidic bonds and presence of numerous reactive hydroxyl groups. The amount of a drug bound to dextran depends upon the degree of activation of the polysaccharide. Dextrans of various molecular weights have been used as drug carriers. There has been no systematic study comparing the different pharmacokinetics of these dextran derivatives containing an identical percentage of bound drugs.

Table 7 gives some examples where several pharmacologically active substances were linked to polysaccharides forming complexes of pharmaceutical interest. They are of remarkable chemical stability and resistence to decomposition at higher

Table 7. Drug-linked polysaccharides

Polysaccharide	Linked compound
Cellulose	albumin
	γ-globulin
	ribonuclease
	trypsin
	antibodies
	concanavalin
Pectin	antigens
	allergens
	toxins
	hormones
Alginic acid	enzymes
	antibodies
	hormones
Dextrans	insulin
	acetylsalicylic acid
	novocaine
	noradrenaline
	amphetamine
	tolbutamide
	ampicilline
	kamamycin
	glutathione
	amylase
	trypsin

temperatures, light, or oxidation, even when the original material is very unstable. From the pharmacokinetic viewpoint, they have half-lives which are much longer than those of the original drugs.

3.5 Ophthalmic Products

The major physical forms are aqueous eyedrop suspensions and ophthalmic ointments. Polysaccharides, such as methyl- and hydroxypropylmethylcellulose, are only used to increase the viscosity of liquid ophthalmic products and the ocular contact time, and further to decrease the drainage rate. The lubrication effect which is important for many patients is a secondary one (Linn and Jones [10], Chrai and Robinson [11]). Direct determination of the ophthalmic bioavailability, which might be influenced by viscosity depressants in the human body is not possible. Fluorescein can be used to study factors affecting bioavailability in the eye (Adler et al. [12]). Over a wide range of viscosity, only a small increase in drug penetration was observed. Ointments for ophthalmic use are normally prepared without polysaccharides.

3.6 Special Dosage Forms: Cyclodextrin Inclusion Complexes

According to the common definition, cyclodextrins are just on the borderline between oligosaccharides and polysaccharides. β-Cyclodextrin, the most important member of this group, comprises seven glucopyranose units (Fig. 1). Cyclodextrins are watersoluble, since all of the free hydroxyl groups are located on the outer surface of the ring; the internal cavity is slightly apolar. One of the typical properties of these amylose-derived products are the inclusion complexes. The cyclic molecules include, only by physical forces without covalent bonding, most molecules having the size of one or two benzene rings. The physicochemical properties of the molecules trapped in the cavity are considerably altered by these complexes, they dissociate easily at physiological conditions and the cyclodextrin-enclosed molecules can thus exert their desired effects. Cyclodextrins can be readily crosslinked, whereby the bead polymers obtained from cyclodextrins possess cavities of well-defined size inside the beads.

It can be expected that these semisynthetic carbohydrate polymers exert quite specific effects in the kinetics of drug release.

Many drugs were shown to form inclusion complexes also with substances which, considering their size, did not seem to be eligible for the complexing reaction.

Lach and Pauli [13] studied the complex formation of β-cyclodextrin with reserpine, cortisone, and testosterone. Kurozumi et al. [14] prepared α- and β-cyclodextrin complexes of non-steroid anti-inflammatory agents with the purpose of obtaining orally administrable pharmaceutical preparations. Indomethacin, as one of the most active substances of this class, was treated with β-cyclodextrin in order to prevent gastric irritation (Szeitli [15]). Frömming et al. [16] showed that several cyclodextrin-drug complexes can readily be compressed into tablets even without additives. These formulas can be used without special precautions. Inclusion complexes of aroma substances can be conveniently applied in order to mask an unpleasant taste of the original drug material. A number of authors studied the inclusion complexes with chemotherapeutic and antitumor agents. Even antibiotics of different chemical

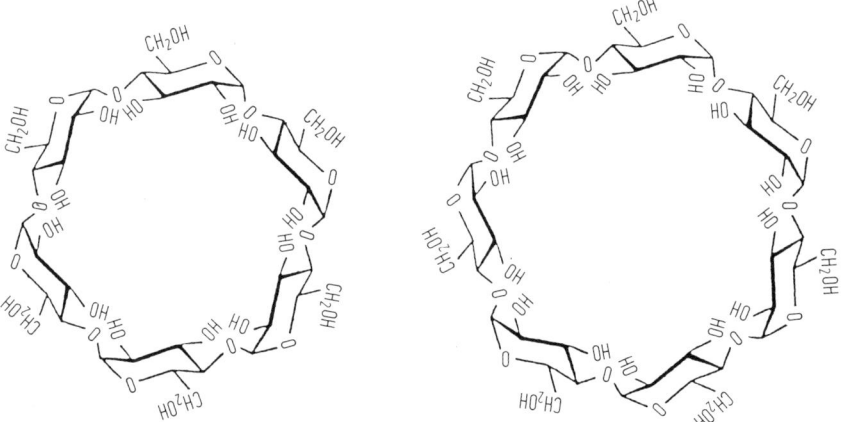

Fig. 1. Structures of α- and β-cyclodextrin

structures gave good results with β-cyclodextrin. These complexes proved to be more stable towards alkali and high temperatures than the original substances (Uekama and Hirayama [17]). A further advantage seems to be the stabilization of active substances in solution as it was shown by Banky [18] in the case of anethole and related substances.

In conclusion one can say that these semi-natural carbohydrate polymeric structures have a wide and prospective field in the pharmaceutical practice.

4 Polysaccharides and their Respective Derivatives from Higher Plants

4.1 Cellulose and Derivatives

The name cellulose is generally given to a group of very closely related substances, rather than to a single entity. Celluloses are anhydrides of β-D-glucose existing as 1,4-linked long chains that are not branched. They consist of several thousand monosaccharide residues with DP values between 4000 and 10000. These chains are crosslinked by hydrogen bonds, thereby supporting the rigid structure of the plant cell wall. Bundles of β-1,4-glucan chains form fibrillar structures, the so-called elementary fibrils. These elementary fibrils are assembled in bundles to give the microfibrils (Fig. 2). Commercial cellulose consists of microfibrillar cellulose with varying chain lengths and crystallinity and different stages of purity depending on the natural source. Cotton cell walls are the purest form of natural occurring cellulose. They still enclose varying amounts of hemicelluloses, pectins, proteins, and lipophilic substances. These must be removed by an appropriate treatment in order to obtain pure cellulose. Other plant celluloses, especially those prepared from wood, can be resolved into a β-cellulose fraction which is soluble in 17.5% NaOH and an alkali-insoluble α-cellulose fraction.

4.1.1 Purified Cotton

Purified cotton consists of the fibers of different cultivated varieties of *Gossypium sp*. The material is freed from adhering impurities, deprived of fatty matter, bleached, and sterilized. The length of cotton fibers is up to about 5 cm, the diameter varies between 9 to 25 μm. A typical cotton fiber is cylindrical when young, but becomes flattened and twisted as it matures. The genuine cellulose wall of the cotton fiber is covered with a waxy cuticule. Delipidation is essential in order to transform the genuine fiber to absorbent cotton wool which is readily wetted by water.

4.1.2 Cellulose Powder, Microcrystalline Cellulose

Cellulose powder is a mechanically shortened cellulosic fiber, whereby the degree of polymerization remains almost intact. Microcrystalline cellulose is a partially purified and depolymerized cellulose, prepared by treating α-cellulose obtained from fibrous plant material with mineral acids. It occurs as a fine white odorless and

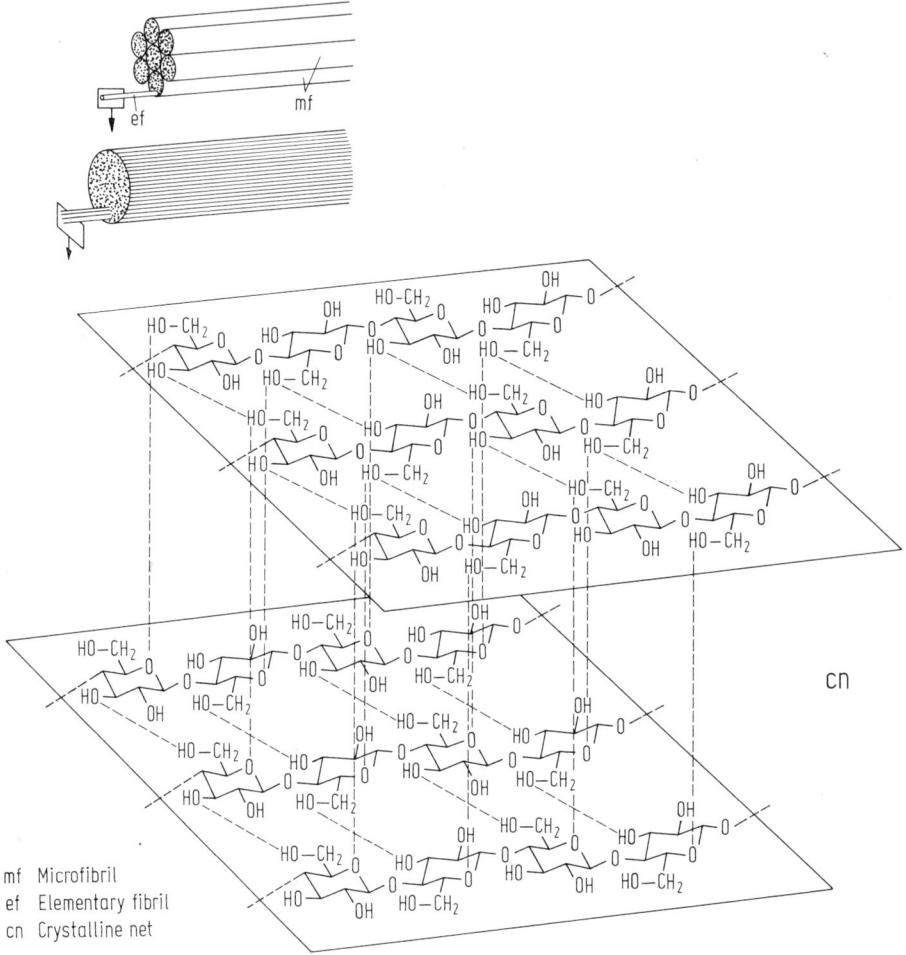

Fig. 2. Molecular and fribrillar structure of cellulose. mf = Microfibril; ef = Elementary fibril; cn = Cristalline net

crystalline powder, which is insoluble in water, dilute alkali, and most organic solvents.

4.1.3 Methylcellulose (Fig. 3)

The methyl ether of cellulose has a methoxyl content which varies between 6 and 33%. Methylcellulose can be prepared by the action of methyl chloride or methyl sulfate on cellulose that has been previously treated with alkali. Purification is accomplished by washing the reaction product with hot water. The degree of methylation can be controlled to yield products with varying viscosities. Seven viscosity types of methylcellulose are currently produced with centipoise values ranging from 10 to 4000. Methylated cellulose of low methoxy content is soluble

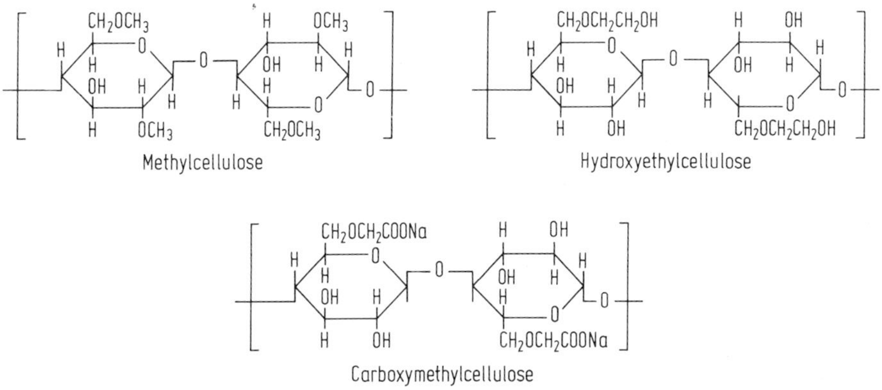

Methylcellulose

Hydroxyethylcellulose

Carboxymethylcellulose

Fig. 3. Some pharmaceutically improtant cellulose derivatives

in cold water but, in contrast to the natural gums, it is insoluble in hot water. Methylcellulose resembles cotton in appearance and is neutral, odorless, tasteless, and inert. It swells in water and produces a clear to opalescent viscous solution. Solutions of methylcellulose are stable in a wide range between pH 2 and 12, with no apparent change in viscosity. Small amounts of ions such as sulfate, phosphate, and carbonate will cause precipitation or coagulation of dissolved methylcellulose. This product is used as a substitute for natural gums or mucilages. Due to its specific viscosity it can be employed as a bulk laxative, further in nose drops, ophthalmic preparations, burn preparations, ointments, and other formulas with defined viscosity.

4.1.4 Hydroxyethylcellulose (Fig. 3)

The ethyl ether of cellulose contains no less than 45% and not more than 50% ethoxy groups. It is available as a free-flowing stable white powder, insoluble in water, glycerol, and other hydrophilic solvents, but soluble in organic solvents such as ethanol, ethylacetate, and chloroform. This cellulose derivative is mainly used as a tablet binder.

4.1.5 Hydroxypropylmethylcellulose

This product contains no less than 19% and no more than 30% methoxy groups and no less than 3% and no more than 12% hydroxypropyl residues. It occurs as a white fibrous granular powder. Hydroxypropylmethylcellulose is employed as a suspending or a thickening agent and tablet excipient. Solutions of this hydrophilic polymer are used as topical protectants or as artificial tears for contact lenses.

4.1.6 Oxidized Cellulose

In oxidized cellulose, part of the terminal primary alcohol groups of the glucose residues are converted to carboxyl groups. The product should contain no less than 16% and no more than 24% carboxyl groups. Products with low carboxyl content have better technical properties. Oxidized cellulose is a fibrous white powder prepared from cotton wool and posesses a slight acid odor.

Oxidized cellulose is very similar to normal cotton but with a defined texture and an acid taste. The material tends to disintegrate on handling. Under the microscope, the fibers are very similar to those of normal absorbent cotton. Oxidized cellulose is used as an absorbable haemostatic in many types of surgery. It is incompatible with penicilline and can not be heat-sterilized.

4.1.7 Cellulose Acetate Phthalate

This cellulose derivative is a partial acetate ester of cellulose, which has been reacted with phthalic anhydride. One carboxyl residue of the phthalic acid is esterified with cellulose acetate. The final product contains about 20% acetate groups and about 35% phthalate groups. In the acid form, it is soluble in organic solvents and insoluble in water, whereas in the salt form it is readily soluble in water. This combination of properties makes it useful in coating of tablets because it is resistant to acids of the stomach but it is readily soluble in the more alkaline environment of the intestinal tract.

4.1.8 Carboxymethylcellulose (CMC) (Fig. 3)

Carboxymethylcellulose is usually prepared as the sodium salt of a polycarboxymethyl ether of cellulose. The procedure permits a control of the number of carboxymethyl groups which are to be introduced. The number of these groups is related to the viscosity of aqueous solutions. Carboxymethylcellulose is available in various viscosity grades ranging from 5 to 2000 centipoises in 1% solutions. The properties of CMC resemble in part those of naturally occurring polysaccharides whose carboxyl groups contribute to the physicochemical characterization of these carbohydrates. Carboxymethylcellulose is used as a suspending or thickening agent, as a tablet excipient, and as a bulk laxative. It is often used in varying proportions together with microcrystalline cellulose to produce suspending agents with different viscosities. CMC is easily dispersed in cold or hot water to form a colloidal solution which is stable at pH 2 to 10.

4.2 Starch and Derivatives

Starch, used in pharmacy, consists of granules separated from the grains or tubers of different higher plants. Starches obtained from different sources may not have identical properties for specific pharmaceutical purposes and, therefore they should not be interchanged in the different formulas.

Many patented processes are used for the isolation of particular starches and the procedure adopted depends on the desired degree of purity and the nature of the compounds from which the starch has been obtained. Cereal starches, for example, have to be separated from cell debries, oil, and soluble proteins.

4.2.1 Properties of Starches

Starch occurs as a white powder which is insoluble in cold water but forms a colloidal solution after boiling with about 15 times its weight of water. The solutions form a translucent jelly on cooling. When starches are heated with water, the granules first swell and then undergo disintegration. The temperatures at which these changes commence and are completed, vary with different starches. Pregelatinized starch, for example, is employed as a tablet excipient. Starch granules also undergo gelatinization after treatment with alkali, concentrated solutions of calcium or zinc chloride, or concentrated solutions of chloral hydrate. Starches of different origin can be identified by microscopic examination. The size, shape, and structure of the starch granules from any particular plant varies only within a well-defined limit so that it is possible to distinguish between starches from different species (Fig. 4). When examined in polarized light, starch granules show birefringence. They appear in the dark field as illuminated objects marked by a dark cross, so that starch behaves like a spherical crystal.

4.2.2 Composition of Starch

Starch granules are composed of two different polysaccharides, amylopectin and amylose; the former constitutes about 80% of the most common starches. Separation of the two components can be achieved by selective precipitation involving the formation of an insoluble complex of amylose with polar organic substances.

Amylose consists of linear chains, whereas amylopectin has a branched structure (Fig. 5). These chemical differences give the two substances different properties which contribute towards the distinctive characteristics of a starch from a particular plant origin.

Amylose, although water soluble, gives an unstable solution which irreversibly precipitates. It is mainly responsible for the deep blue coloration given by starch and iodine. Solutions of amylopectin are relatively stable. The iodine-binding capacity, on the other hand, is very low. A small amount of covalently bound phosphate normally appears with starch but its exact location within the molecule is not known.

Starch is extensively used due to its adsorbing properties. In dissolved form, it is used as a skin emollient and as an antidote for iodine poisoning. Other applications include the use as a tablet filler and binder and disintegrant. Sterilized starch is used as a lubricant for surgeon gloves. Unlike talc, it is completely adsorbed by body tissues. Soluble starch is prepared by treating commercial potato starch with dilute hydrochloric acid until it forms an almost clear solution in hot water.

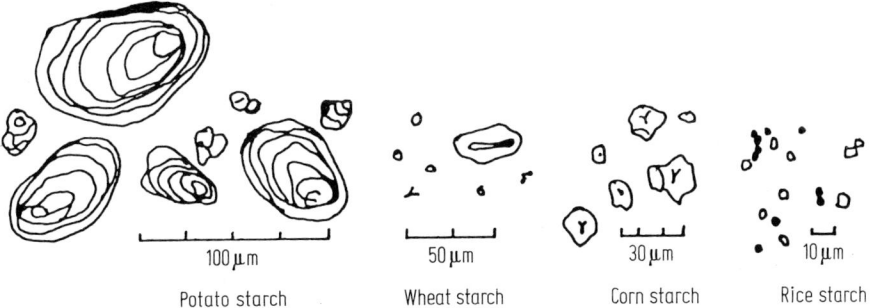

Fig. 4. Pharmaceutically used starches

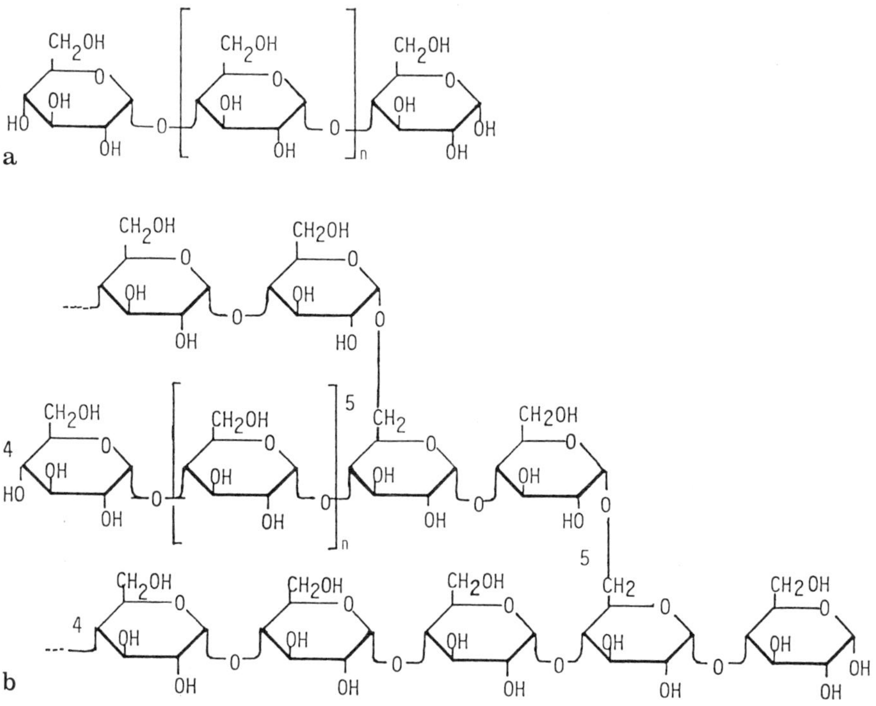

Fig. 5a. Repeating unit of amylose.   b. Repeating unit of amylopectin

4.2.3 Dextrin

Dextrins (white dextrin) are prepared by heating starch, which has been moistened with a small quantity of dilute nitric acid and dried at 110–120 °C. Dextrin may contain up to 15% soluble starch. Yellow dextrins are more completely decomposed and, unlike the white variety, they contain appreciable amounts of maltose which may be detected and estimated by their reducing power.

4.3 Gums and Mucilages

Gums are natural plant hydrocolloids that may be classified as ionic or nonionic polysaccharides or salts of polysaccharides. They are amorphous substances that are frequently produced in higher plants as a protective after injury. Gums are typically heteropolysaccharides in their chemical composition. Upon hydrolysis, L-arabinose, D-galactose, D-glucose, D-mannose, D-xylose, and D-galacturonic acid and D-glucuronic acid are the most frequently observed components (Fig. 6). The uronic acids may form salts with calcium, magnesium, and other divalent cations. Methylester and sulfate-ester substituents further modify the hydrophilic properties of some natural polysaccharides. Some efforts have been made to distinguish between

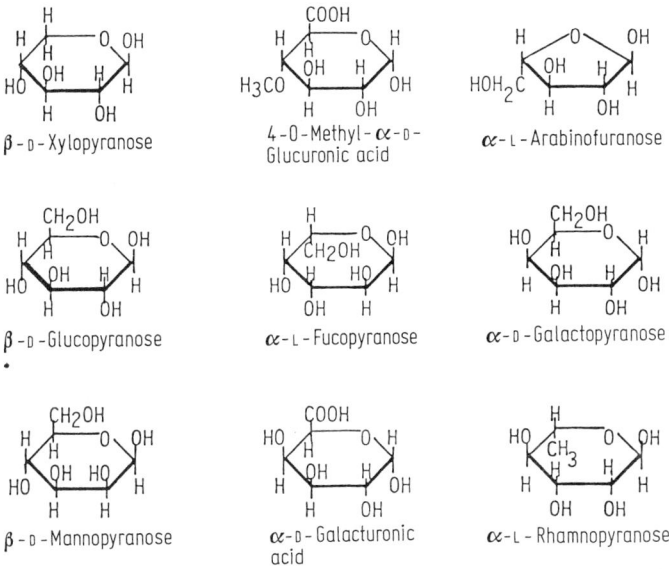

Fig. 6. Sugar components of gums

mucilages and gums. Gums readily dissolve in water while most mucilages form slimy masses. Mucilages are physiological plant products and gums are pathological products. However, these classifications have not been very successful. Exact knowledge about the chemical nature of these polysaccharides is increasing and associated with the practical usefulness of the physical and chemical properties.

Plant hydrocolloids may be linear or branched, they may have acidic or neutral characteristics and show a wide range of molecular dimensions.

Gums consisting of linear polymers are less soluble and give solutions with greater viscosity than those with branched structures (Tab. 8). These features are related to the increased possibility for good alignment and considerable intermolecular hydrogen bonding among linear molecules. This tendency for intermolecular association also explains why solutions of linear polysaccharides are less stable and tend to precipitate especially at lower temperature, compared to solutions of branched polysaccharides. In linear polymers containing uronic acids, the intermolecular associations are reduced and these products give more stable solutions.

Branched hydrocolloids form gels rather than viscous solutions at higher concentrations. They tend to be sticky when moistened; a common feature which is of advantage for adhesive purposes. They rehydrate more readily than linear hydrocolloids, which is an important property for drug formulas that must be reconstituted immediately before use.

Several plant exudates (gums and mucilages) have been used for pharmaceutical purposes and they still find significant application. However, the production of gums is laborious and expensive and their use will probably continue to decline.

Table 8. Viscosity of polysaccharides in solution (cP) (modified after Belitz and Grosch [19])

Conc. %	Gum arabic	Tragacanth	Carrageenan	Sodium alginate	Guaran
1	—	54	57	214	3025
3	—	10605	4411	29400	111150
5	7.3	111000	51425	—	510000
10	16.5	—	—	—	—
50	4162	—	—	—	—

4.3.1 Acacia Gum (Gum Arabic, Acaciae Gummi)

Acacia gum is the dried exudate from branches of *Acacia senegal* or other related African species of *Acacia*. It is commonly known under the name of gum arabic. This name probably reflects its extensive use by the early Arabian physicians. Acacia plants grow in Sudan and Senegal, where the trees are tapped by making a transverse incision in the bark or peeling off the bark. In about two weeks, tears of gum, which have a diameter from 2 to 7 cm, form on the exposed surface, they dry and are collected. The average annual yield of gum per tree ranges from 900 to 2000 g. The formation of the gum may also be caused by bacterial infection or by the action of enzymes. In most areas, the gum is exposed to the sun and bleached, whereby minute cracks are formed, giving the gum an opaque appearance. Gum arabic is a mixture of several related polysaccharides with a molecular weight ranging from 2.6×10^5 to 1.2×10^6. L-arabinose, L-rhamnose, D-galactose, and D-glucuronic acid have been demonstrated to be the molecular building units which form the highly branched polysaccharide (Fig. 7). The main polysaccharide is composed of β-D-galactopyranosyl residues connected by 1.3-, or 1,6-linkages. At

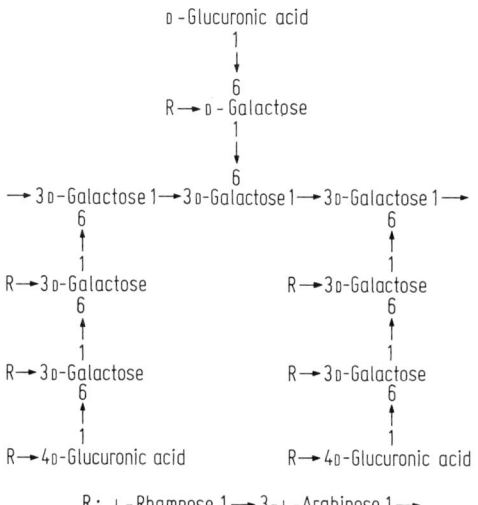

Fig. 7. Basic structure of gum arabic

the C-6 atoms they carry side chains of galactose residues onto which arabinose and rhamnose units are bonded. The genuine polysaccharide is neutral or slightly acid. Counterions have been shown to be calcium, magnesium, or potassium. After dissolving the polysaccharide in 0.1 N HCl and precipitating with ethanol, the free acid can be obtained. Acacia gum is almost completely soluble in an equal weight of water, the dissolution takes place rather slowly. The solution is slightly acid and becomes more so on keeping, especially if hot water is used to make the solution.

A 10% solution gives no precipitate with a diluted lead acetate solution (distinction from tragacanth and agar), it should give no color with iodine (absence of starch and dextrin), and should give no reaction for tannin with ferric chloride. The viscous solution produces a blue color when treated with benzidine, and with a few drops of hydrogen peroxide indicating the presence of peroxidases (distinction from tragacanth). The good solubility of arabic gum allows to prepare solutions up to a concentration of 50%. The viscosity of the solution increases only at high concentrations compared to other polysaccharide solutions, which are already highly viscous at low concentrations.

Solutions of acacia gum have a low viscosity and good stability over the range from pH 2 to 10. These properties of the gum are the base for the use as an excellent emulsifying agent. Acacia gum is also employed as a suspending agent; it demonstrates useful demulcent and emulgent properties and finds application as an adhesive and a binder in tablet granulations.

4.3.2 Tragacanth Gum

Tragacanth is a gum which is obtained by incision of the stems of various species of *Astragalus*. The main gum-yielding species are thorny shrubs found in the mountains of Asia minor, Syria, Armenia, Kurdistan, Irak, and Iran.

When the plant is injured, the cell walls of the pith and of the medullary rays are gradually transformed into gum. The gum absorbs water and produces internal pressure within the stems, thus forcing the gum to the surface through the incision caused by the injury. It gradually hardens owing to the evaporation of water.

Tragacanth consists of a water-soluble fraction known as tragacanthin and a water-insoluble fraction known as bassorin. The soluble fraction (tragacanthin) is a complex mixture of acid polysaccharides with the following molecular constituents: D-galacturonic acid, D-galactose, L-fucose, D-xylose, and L-arabinose. In the insoluble fraction, the carboxyl groups of the uronic acid are mainly esterified with methanol. It has been shown that the best grades of gum contain the least amount of tragacanthin. The molecular weight ranges between 8×10^5 and 1.5×10^6. The molecules seem to be very stretched, which is a prerequisite for solutions of high viscosity.

Tragacanth is pharmaceutically employed as a suspending agent for insoluble powders, further as an emulsifying agent for oils and resins, and as an adhesive. Tragacanth is the most resistent of the hydrocolloids to acid hydrolysis and is therefore preferred for any use under highly acidic conditions. In some formulas, it has been employed as a binding agent for pills and tablets and also as a demulcent.

4.3.3 Sterculia Gum (Karaya Gum, Indian Tragacanth)

Sterculia gum is the dried exudate obtained from the tree *Sterculia urens*. It is produced in India, Pakistan, and to a small extent in Africa. The gum has only relatively recently been introduced having been generally regarded as an inferior substitute for tragacanth in the beginning of this century. Today, it has been shown to be superior in some physical and chemical properties. Sterculia gum occurs in irregular translucent masses differing from tragacanth by containing no starch. In water, Sterculia gum has a low solubility but swells to many times its original volume.

Acid hydrolysis of Sterculia gum gives rise to the formation of D-galactose, L-rhamnose, and D-galacturonic acid. The galacturonic acid units are the branching points of the complex molecule.

This gum is used as a bulk laxative, as an agent for forming emulsions and suspensions, and as a dental and stoma appliance adhesive.

4.3.4 Plantago Seed (Psyllium-Flea Seed)

Plantago seeds are the dried ripe seeds of *Plantago psyllium* or *Plantago indica*, known as Spanish or French psyllium sed. All seeds contain between 10–30% mucilage in the epidermal part of the testa. The seeds may be evaluated by measuring the swelling factor, i.e., the volume of mucilage produced from 1 g of seeds. The typical swelling factor for *Plantago psyllium* was established as 12.7, for *Plantago arenaria* 14.5, and 10.5 for Plantago ovata. The hydrocolloid can be separated into an acidic and a neutral polysaccharide fraction. Upon hydrolysis, L-arabinose, D-galactose, D-galacturonic acid, L-rhamnose, and D-xylose are obtained.

Solutions of purified *Plantago* mucilages are thixotropic: the viscosity decreases as the shear rate increases. Plantago seeds are used in the treatment of constipation. Its action is caused by the swelling of the mucilagious seed coat, thus giving bulk and lubrication. The seeds should always be taken with a consideruble amount of water. Psyllium seeds are used as ingredients in a number of laxative formulas, often combined with various laxative agents.

4.3.5 Quince Seed

Quince seed is the ripe seed of *Cydonia vulgaris*. In the seed coat there is a mucilaginous outer layer making up approximately 20% of the seed weight. The mucilage is composed of cellulose fibrils suspended in the more soluble polysaccharide fraction, consisting of L-arabinose and a mixture of aldobiuronic acids. The mucilage forms viscous solutions with thixotropic properties and is used as an emulsifying agent.

4.4 Pectic Substances

Pectin and related substances are widespread in the plant kingdom. They are normally obtained from the dilute acid extract of the inner portion of citrus peels or from apple pomace. The extraction of pectin is carried out at elevated temperatures at a controlled pH. The process must be very carefully checked in order to avoid a possible hydrolysis of glycosidic and ester linkages. The crude extract is concen-

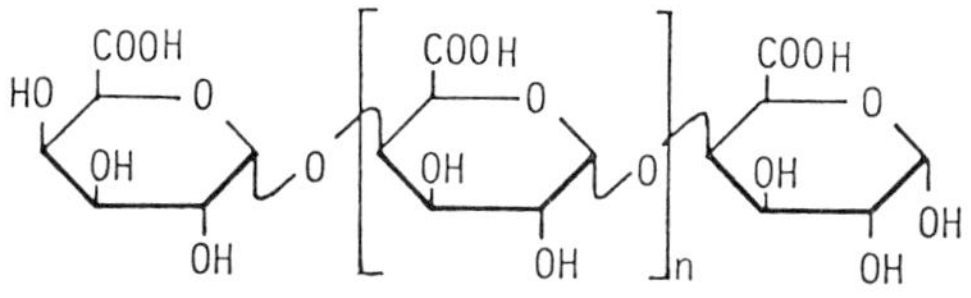

Fig. 8. General structure of pectic acid

trated and purified by precipitation with ions forming insoluble salts with the uronic residues. Isopropanol and ethanol also can be used as precipitating agents.

Pectic acid is a hydrophylic colloid consisting of partly methoxylated 1,4-linked polygalacturonic acid (Fig. 8). In protopectin, lateral chains covalently bound to the main chain are composed of D-galactose, L-arabinose, and L-rhamnose. The molecular weight of pectin ranges from 100 000 to 250 000. Pharmaceutical pectin is pure pectic acid to which no additions have to be made. At pH 3, pectin solutions form thermoreversible gels in the presence of calcium ions. The gel-forming capacity increases with increasing molecular weight. Low-esterified pectins require low pH values and calcium ions to form gels, while no further sugars or organic acids are needed. In contrast, highly esterified pectins require the addition of sugars and organic acids in order to gelatinize.

Commercial pectin is a fine, yellowish white, almost odorless powder with a mucilaginous taste. It is completely soluble in twenty parts of water giving an acidic, viscous, and opalescent solution.

Pectin is mainly used as a suspending agent and is an ingredient in many antidiarrheal formulas. As a colloidal solution, it conjugates toxins and enhances the physiological functions of the intestinal system through its physical and chemical properties.

4.5 Galactomannans

Galactomannans are found in a great number of plants, many of which are of considerable pharmaceutical interest. The importance of galactomannans and their biochemistry has been reviewed by Whistler [20] and by Dey [21].

The common molecular structure consists of 1,4-linked β-D-mannopyranosyl units to which varying amounts of α-D-galactopyranosyl groups are joined (Fig. 9).

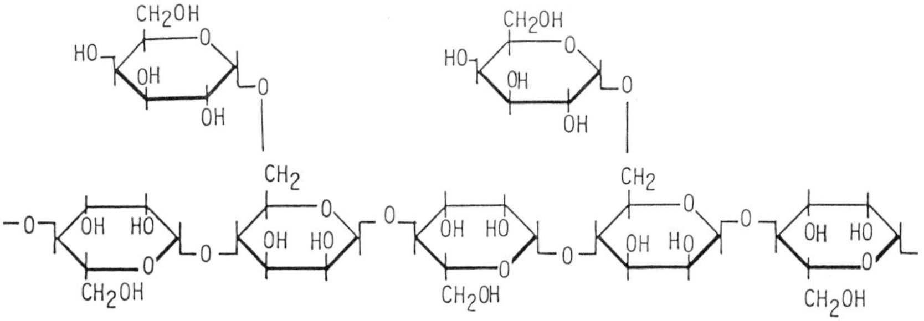

Fig. 9. General structure of Galactomannan

The linear, but highly branched, molecular structure of the galactomannans is the reason for some specific properties which are quite different from those of the unbranched cellulose-like and water-insoluble mannans and glucomannans. Galactomannans are hydrated in cold water and give stable solutions even in acidic formulas. The interactions of galactomannans with other polysaccharides are the base of a variety of industrial applications.

4.5.1 Galactomannan Sources

Guar gum or guaran is the powdered endosperm of the seeds of *Cyamopsis tetragonolobus*, an annual plant cultivated in dry climates. The use of this gum is expanding rapidly and there is almost no limit of the amount of gum that can be produced. Locust bean gum is obtained from the powdered endosperm of the seeds of *Ceratonia siliqua*, a tree native to the Mediterranean region. The slow growth of the tree restricts the prospects for increasing the supply of the gum to meet expanding demands. The structural distinction from guar gum consists in the lower frequency of galactose residues on the linear mannose chain, where only every 4th or 5th mannose residue is substituted.

Galactomannanas are polysaccharides with manyfold applications. They are used as bulk-forming laxative and thickening agents, as tablet binders, and as disintegrators in pharmaceuticals. More recently, spray inbeddings for antibiotics and several vitamins have been shown to be an area where glactomannans can successfully be employed (Nürnberg [22]). Recently, galactomannans have been carboxymethylated. Aqueous solutions of these galactomannan derivatives show pseudoplastic fluidity in contrast to the non-carboxymethylated products. They further demonstrate a significant increase in viscosity at higher pH values (Nürnberg and Pritting [23]).

It has been known for some time that pectin reduces the level of cholesterol in the blood, and, more recently, a similar effect has been described for several galactomannans. They are also very useful in the treatment of diabetes as it has been shown by Jenkins et al. [24].

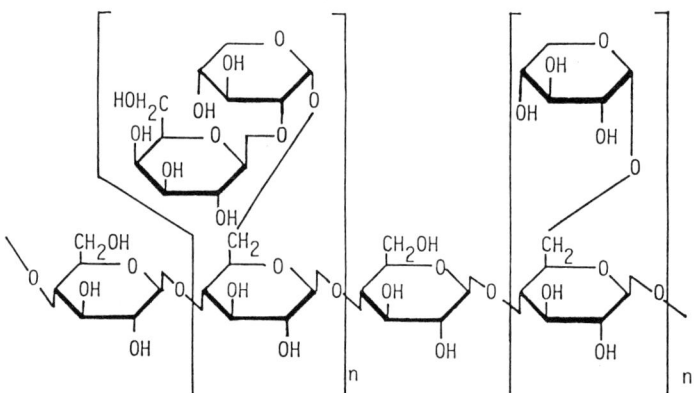

Fig. 10. General structure of amyloid (Xyloglucan)

4.6 Xyloglucans

The seeds of *Tamarindus indica* contain polysaccharides which can be extracted with hot water and purified by precipitation with alcohol. The polysaccharide is composed of D-galactose, D-xylose, and D-glucose (Fig. 10). These xyloglucans have been named amyloids since they give a color reaction with iodine similar to starch. The xyloglucans from *Tamarindus* are gel-forming polysaccharides which are stable in a wide pH range. This type of polysaccharide is mainly used in food industry, but, in some cases, it has been pharmaceutically employed as a substitute for pectin.

5 Polysaccharides from Algae

The three major divisions of marine algae, i.e., *Phaeophytae* (brown algae), *Rhodophytae* (red algae), and *Chlorophytae* (green algae) produce a large variety of different cell walls and yield polysaccharides with very interesting physical and chemical properties. They are characterized by the occurrence of some sugar constituents which have not been found in land-plant cell walls.

5.1 Alginic Acid

Large quantities of brown algae are collected in Japan, USA, Canada, and Scotland having an alginic acid content of about 20–40%. This polysaccharide is the main cell wall constituent of brown algae occurring as a mixture of the free acid and corresponding salts. Alginic acid is extracted from the plant material and converted into the water-soluble sodium alginate. This product normally contains small amounts of sulfated polysaccharides and protein.

It can be further purified by precipitation with calcium salts. During extraction and purification, some degradation of the polymer takes place and the preparations show varying degrees of polymerization ranging from 1000 to 10000.

Alginic acid is a linear polymer composed of β-1,4-D-mannuronic and L-gulucuronic acid; the proportions of the two units are not constant in the alginates from different algae species (Fig. 11).

Sodium alginate is a nearly odorless and tasteless powder which is readily soluble in water, forming a viscous colloidal solution. Free alginic acid is insoluble in cold

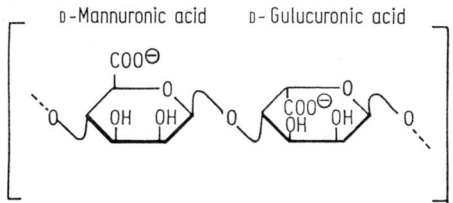

Fig. 11. Repeating unit of alginic acid

water, slightly soluble in hot water, but it swells and adsorbs many times its own weight.

Because of the greater chemical reactivity, the sodium alginates have certain advantages over starch, agar, pectin, and other plant gums. Free alginic acid is used as a tablet binder and as a thickening agent. The alginates are applied as stabilizing, thickening, emulsifying, deflocculating, and gel-forming agents. They are commonly used in creams, ointments, pastes, jellies, and tablets. Mainly, calcium alginate is used for a number of gelation purposes including the formation of a gel for preparing dental impressions.

5.2 Agar

This polysaccharide is extracted from different red algae which grow along the coastal regions of eastern Asia, USA, and Europe. For their preparation, the fresh seaweed is washed and digested with dilute acid and water. The extract is cooled and freed from the solvents by a freeze-drying process. Commercial agar usually occurs in bundles consisting of thin agglutinated strips or in cut flaked or granulated form. The product is odorless with a yellowish color. Agar is insoluble in cold water, but if one part of agar is boiled for 10 min with 60 times its weight of water, it gives a firm gel after cooling. The predominating building blocks of agar are D-galactose and 3,6-anhydro-L-galactose units which are alternately linked by 1,4- and 1,3-linkages (Fig. 12).

Agar can be ressolved into two major fractions, i.e., agarose and agaropectin. Agarose shows a low esterification degree with sulfuric acid, whereas the sulfate ester content of agaropectin is much higher.

Agar is used as a suspending and emulsifying agent, as matrix substance for suppositories, as surgical lubricant, as a tablet excipient and disintegrant. After hydration, it can be employed as a bulk laxative by stimulating the intestinal peristaltic. In bacteriology it is extensively used as a gel for culture media.

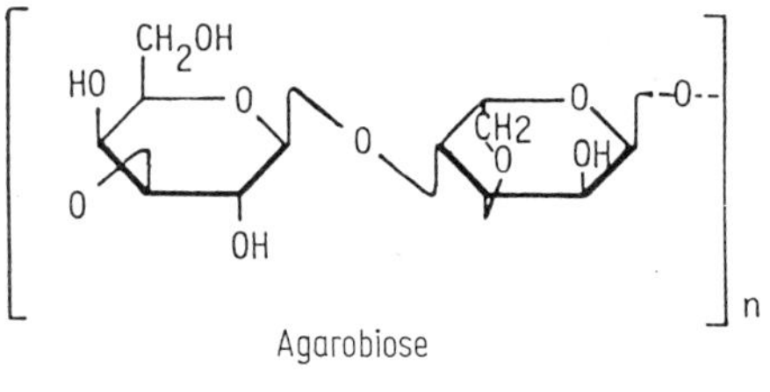

Agarobiose

Fig. 12. Repeating unit of agarose

5.3 Carrageenan

The term carrageenan covers a whole range of sulfated polysaccharides extracted from various red seaweeds. Two major carrageenan fractions have been isolated, namely ϰ-carrageenan and λ-carrageenan, which differ in their gel-forming capacities. Both compounds are composed of galactose residues which are sulfated and alternately linked in 1,3 and 1,4 positions (Figs. 13, 14). ϰ-Carrageenan can be ob-

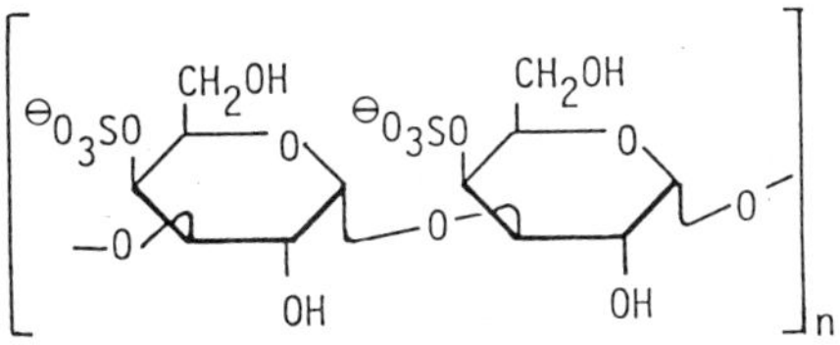

Fig. 13. Repeating unit of λ-carrageenan

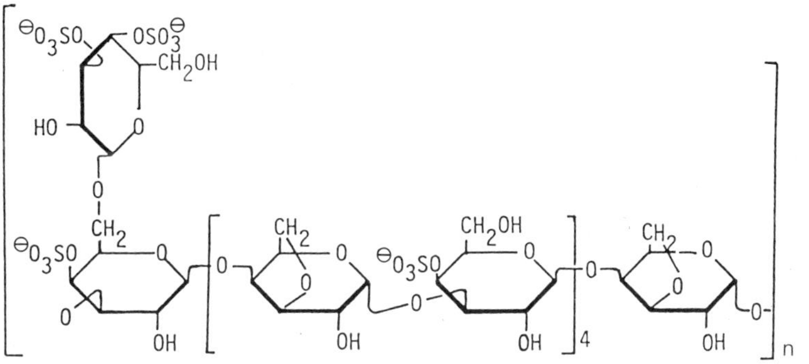

Fig. 14. Repeating unit of ϰ-carrageenan

tained from water extracts of *Chondrus sp.* or *Gigartina sp.* by precipitating with potassium ions. The material left in solution is the λ-carrageenan.

In aqueous solutions at elevated temperature, carrageenans exist as random coils. On cooling, a three-dimensional polymer network builds up, in which double helices of ϰ-carrageenan form junction points of the polymer chains.

In combination with galactomannans, ϰ-carrageenan shows an unusual synergism, which is marked by an enhancement of the gel strength. This phenomenon is believed to be due to a carrageenan-galactomannan interaction.

Mainly the gel-forming properties of ϰ-carrageenan are used for pharmaceutical purposes. Due to the high viscosity of the λ-carrageenan, it is employed as an emulsifying and stabilizing agent.

5.4 Furcelleran

Furcelleran is extracted from the red seaweed *Furcelleria fastigiata*. This polysaccharide is extracted after an alkali pretreatment with hot water. The extract is treated with potassium ions, whereby the gel precipitates. Furcelleran is composed of D-galactose and 3,6-anhydro-D-galactose and has sulfate ester groups on both sugar components. The structure of furcelleran is comparable with that of ϰ-carrageenan. It differs primarily in the amount of sulfate esters. Gel formation and gel textures are also similar. Furcelleran is used as a gelating and suspending agent.

6 Polysaccharides from Bacteria

As with higher organisms, a common feature of bacteria is the production of extracellular polysaccharides during growth. Within the last 20 years, the large-scale production of microbial biopolymers has become feasible, and mainly two microbial products, i.e., xanthan and dextran are widely used in the pharmaceutical industry today.

6.1 Xanthan

This product is a high-molecular-weight microbial gum prepared by the action of *Xanthomonas campestris* on a glucose medium. The exocellular gum is recovered from the fermentation broth by precipitating with isopropanol. Gums with various genetically controlled properties and molecular weights are available, mostly as the sodium-, postassium-, or calcium salts. The structure of xanthan gum (Fig. 15) consists of a main chain of β-1,4-linked D-glucose units and side chains containing D-mannose and D-glucuronic acid residues.

Xanthan gum shows a good solubility in water, giving a highly viscous solution with a pseudoplastic appearance and a temperature independent viscosity. Xanthan gum is used in pharmaceuticals for its excellent emulsifying and suspending properties. The pseudoplastic properties of this gum enables tooth pastes and ointments both to hold their shape and to spread readily.

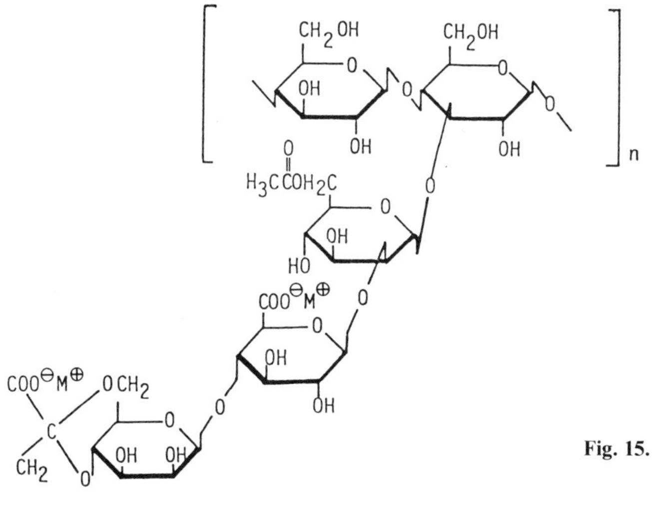

Fig. 15. Repeating unit of xanthan

6.2 Dextran

The name dextran is given to a large class of exocellular bacterial glucans. Although each bacterial strain produces a unique glucan, a common feature of all dextrans is a large percentage of α-1,6-linkages with smaller proportions of α-1,2-, 1,3-, or

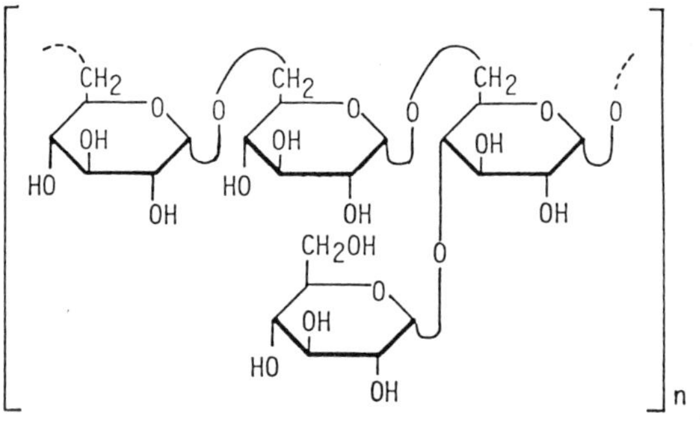

Fig. 16. Repeating unit of dextran

1,4-linkages resulting in a highly branched molecule. In general, dextrans having the highest content of 1,6-linkages are the most soluble (Fig. 16).

Dextran gels used as molecular sieves are formed by crosslinking dextrans with epichlorhydrin to give a semisynthetic polysaccharide with a well-defined pore size (Sephadex®).

The molecular-sieve dextran gels are widely used in chemistry, biochemistry, and pharmacy for the analytical and preparative separation of metabolites and other biological products.

Commercial dextrans have average molecular weights ranging from 40000 to 110000. Dextran solutions are inert and do not affect the cell viability. Because of these properties, dextrans are used as plasma expanders to maintain or to replace blood volume. For this purpose, sterile and pyrogen-free solutions of dextran with an average molecular weight (M) of 40000, 70000, or 100000 are employed, depending upon the clinical needs. Dextran with a molecular weight of 40000 or less is completely excreted in the urine within 48 h. Above M = 40000, dextran remains in the blood for prolonged periods of time. Thus, only 20% of dextran of M = 40000 is found in the blood 6 h after injection, while 40% of dextran of M = 70000 is present in the blood within 24 h. Dextran of M = 110000 can remain in the blood for two to three days.

The varying rates of dextran clearance have led to the idea to use these inert polysaccharides as drug carriers (see Sect. 3.4).

7 Polysaccharides as Antitumor Agents

Since several decades the observation has been made that tumor growth may regress following bacterial infection. It was supposed that polysaccharides of bacterial origin were responsible for this phenomenon. Extensive studies have been carried out on polysaccharides with a possible antitumor activity. Today, many studies are in progress, and the role of specific polysaccharides influencing the immune system is being debated for several years (Wagner [25]).

The earliest antitumor-polysaccharide known was isolated in 1943 from *Serratia marcescens* and became known as Shear's polysaccharide (Shear et al. [26]). Later on, a whole steries of polysaccharides of bacterial origin was tested and it became obvious that the ability to cause tumor hemorrhage decreased, when the carbohydrates were sequentially removed from the polysaccharide chains (Malkiel and Harris [27]). The polysaccharides from *E. coli* and other microorganisms were further shown to be active against Sarkoma 180 and Ehrlich carcinoma (Watanabe [28]).

In recent years, many fungi were examined for potent antitumor polysaccharides and a great number of different carbohydrate polymers could be isolated and shown to have a quite distinct antitumor activity. The most active ones have β-1,3- and 1,6-linkages and glucose as the main structural monomer.

Belkin et al. [29] were first to examine various polysaccharide fractions from higher plants for their antitumor activity. They could demonstrate that many of these fractions produced haemorrhagic necrosis in different tumor types. In most cases, the polysaccharides were injected intraperitoneally into mice carrying Sarkoma 37 ascites tumor. The result was a progressive increase in cell volume and in cytoplasmic vacuolization. Osswald [30] found that tragacanth, gum arabic, and CMC reduced tumor cells in Ehrlich ascites carcinoma in female NMRE mice. The effect depended upon the dose, the route of injection, and the molecular size of the polysaccharides administered.

However, oral administration of these polysaccharides had no effect. Hemicelluloses from different higher plants caused regression of solid Sarkoma 180 in mice but not of ascites tumor. Arabinoglucuronoxylan isolated from wheat-straw hemicellulose was completely devoid of activity, whereas arabinoglucoxylan was highly active (Nakahara [31]).

7.1 Polysaccharide Structure and Antitumor Activity

As it has been shown in many cases, polysaccharides, even if they are composed of only one kind of sugar, can possess complex structures. They also form secondary structures depending upon the conformation of the component sugar residues, the chain length, and the chain interactions. The situation is even more complicated when polysaccharides are composed of two or more kinds of monosaccharide residues.

All polysaccharides of different origin with pronounced antitumor activity differ greatly in the composition of their chemical and physical structures. Antitumor activity has been shown to be related to a wide range of different polysaccharides starting from homopolymers such as glucans and mannans up to highly complex glucans such as hemicelluloses, gums, and other complex polysaccharides.

Whistler et al. [32] concluded that mostly the soluble glucans are active antitumor agents, mainly if they are linear. Polysaccharides, which are hydrolyzed by endogeneous enzymes of the intestinal system, e.g., glycogen, starch, and dextrins, are completely inactive. β-1,3-Glucans are more active than β-1,6-glucans; β-1,4-glucan (cellulose) is inactive, while methylcellulose and carboxymethylcellulose are active. Native dextrans are inactive unless derivatized with diethylaminoethyl groups. The mode by which several polysaccharides show pronounced antitumor activity is not entirely clear. Certain bacterial polysaccharides may directly attack tumors, but polysaccharides from fungi and higher plants could not be shown to have a direct action on tumor cells.

In some cases with crude preparations from higher plants, it has been shown that the cell volume increased highly, and neoplasmic cell vacuolization was observed. As a result of the administration of polysaccharides, the membranes of Ascites cells showed increased permeability to solutes and, consequently, the cell inhibed more water and swelled.

The most likely effect of polysaccharides upon tumor cells might be related to the immune response.

Wagner et al. [33] have shown that two distinct polysaccharide fractions from *Echinacea purpurea* exhibited pronounced activities characterized by a high rate of phagocytosis stimulation. One polysaccharide was shown to be a heteroxylan of molecular weight 35 000 and an arabinorhamnogalactan of molecular weight 450 000. The main characteristics of these polysaccharides from *Echinacea* were the optimal solubility in water, the high content of uronic acids, and the very complex structure.

However, as it has been concluded by Whistler et al. [32], all the polysaccharides so far examined are active in transplanted tumors but do not seem to be useful yet for clinical trials in human cancer therapy. The search for new polysaccharides which are more strongly active antitumor agents than those so far discovered stands at a challenge for further studies on antitumor polysaccharides.

8 References

1. Sandford, P. A., Baired, J., in: The Polysaccharides. Vol. 2, p. 441, G. O. Aspinall (Ed.), New York, London: Academic Press, 1983
2. Pedersen, J. K., in: Polysaccharides in Foods, p. 219, London: Butterworths 1979
3. Nürnberg, E., Rettig, E.: Pharm. Ind. *36*, 194 (1974)
4. Nürnberg, E., Bleimüller, G.: Pharm. Ind. *42*, 1292 (1980)
5. Balley, A., Campbell, D. H.: J. Am. Chem. Soc. *85*, 487 (1963)
6. Lilly, M., Money, C., Hornby, W., Crook, E. M.: Biochem. J. *95*, 45 (1965)
7. Haimowich, J. et al.: Nature *214*, 1369 (1967)
8. Molteni, L., Scrollini, F.: Eur. J. Med. Chem. *9*, 618 (1974)
9. Marshall, J. J., Rabinowitz, M. L.: J. Biol. Chem. *251*, 1081 (1976)
10. Linn, M. L., Jones, L. T.: Amer. J. Ophtal. *65*, 76 (1968)
11. Chrai, S. S., Robinson, J. R.: J. Pharm. Sci. *63*, 1218 (1974)
12. Adler, C. A., Maurice, D. D., Peterson, M. E.: Exp. Eye Res. *11*, 34 (1971)
13. Lach, J. L., Pauli, W. A.: Pharm. Sci. *55*, 32 (1966)
14. Kurozumi, M., Nambu, T., Nagai, T.: Chem. Pharm. Bull. *23*, 3062 (1975)
15. Szeitli, J.: Cyclodextrins and their Inclusion Complexes, Budapest, Akademia Kiado 1982
16. Frömming, K. H., Weyer, J.: Arch. Pharm. *305*, 290 (1972)
17. Uekama, K., Hirayama, F.: Chem. Pharm. Bull. *26*, 1195 (1978)
18. Banky, E., Thesis, Ph. D.: Debrecen (1975)
19. Belitz, H. D., Grosch, W.: Lehrbuch der Lebensmittelchemie, p. 243, Berlin, Heidelberg, New York: Springer Verlag 1982
20. Whistler, L.: Industrial Gums, 2nd ed. New York: Academic Press 1973
21. Dey, P. M.: Biochemistry of plant galactomannans. Adv. Carbohydr. Chem. Biochem. *35*, 450 (1978)
22. Nürnberg, E.: Dtsch. Apoth. Ztg. *109*, 1103 (1969)
23. Nürnberg, E., Pritting, D.: Dtsch. Apoth. Ztg. *123*, 302 (1983)

24. Jenkins, D. J. A., Leeds, A. R., Newton, C.: Lancet 779 (1977)
25. Wagner, H.: Fortschr. Arzneimittelforschung 133 (1984)
26. Shear, M., Turner, F. C.: J. Nat. Cancer Inst. *4*, 81 (1943)
27. Malkiel, S., Harris, B. J.: Cancer Res. *21*, 1461 (1961)
28. Watanabe, T.: Jap. J. Expt. Med. *36*, 453 (1966)
29. Belkin, M. et al.: Cancer Res. *19*, 1050 (1959)
30. Osswald, H.: Arzneimittelforschung *18*, 1495 (1968)
31. Nakahasa, W., Tokuzen, R., Whistler, R. L.: Nature *216*, 347 (1967)
32. Whistler, L., Bushway, A., Singh, P. P.: Adv. Carbohydr. Chem. Biochem. *32*, 235 (1976)
33. Wagner, A. et al.: Drug. Res. *34*, 659 (1984)

Editor: K. Dušek
Received June 20, 1985

Thermomechanics of Polymers

Yu. K. Godovsky
Karpov Institute of Physical Chemistry, Moscow B-120/USSR

This article reviews recent developments in polymer thermomechanics both in theory and experiment. The first section is concerned with theories of thermomechanics of polymers both in rubbery and solid (glassy and crystalline) states with special emphasis on relationships following from the thermomechanical equations of state. In the second section, some of the methods of thermomechanical measurements are briefly described. The third section deals with the thermomechanics of molecular networks and rubberlike materials including such technically important materials as filled rubbers and block and graft copolymers. Some recent data on thermomechanical behaviour of bioelastomers are also described. In the fourth section, thermomechanics of solid polymers both in undrawn and drawn states are discussed with a special focus on the molecular and structural interpretation of thermomechanical experiments. The concluding remarks stress the progress in the understanding of the thermomechanical properties of polymers.

1 Introduction 33

2 Thermomechanics of Elastic Systems 34
  2.1 Phenomenological Aspects 34
  2.2 Thermomechanics of Quasi-isotropic Hookean Solids 36
    2.2.1 Uniform (Volume) Dilatation and Compression 36
    2.2.2 Simple Elongation and Compression 37
    2.2.3 Thermal Expansivity 38
    2.2.4 Shear. Torsion 39
  2.3 Thermomechanics of Molecular Networks 39
    2.3.1 Thermomechanical Equations of State Based on Statistical Theories 40
      2.3.1.1 Thermomechanics of Gaussian Networks 40
        2.3.1.1.1 Uniaxial Deformation 40
        2.3.1.1.2 Torsion 46
      2.3.1.2 Thermomechanics of Non-Gaussian Networks 47
    2.3.2 Phenomenological Equations of State 48
    2.3.3 The New Developments of the Theory of Elasticity of Polymer Networks 51

3 Experimental Methods 54
  3.1 Measurements of the Temperature Changes (Isoentropic Measurements) 55
  3.2 Temperature Dependence of Stresses (Isometric Measurements) . . . 55
  3.3 Deformation Calorimetry (Isothermal Measurements) 56

4 Thermomechanics of Molecular Networks and Rubberlike Materials 57
  4.1 Intrachain Energy Effects 57
  4.2 Interchain Effects 61
    4.2.1 Thermomechanical Inversions 62
    4.2.2 Strain-Induced Volume Dilation 63
  4.3 Thermomechanics at Large Deformations 66
  4.4 Thermoelasticity of Liquid Crystalline Networks 67
  4.5 Thermomechanical Behaviour of Rubberlike Materials 68
    4.5.1 Stress Softening: Energetics and Mechanisms 69
    4.5.2 Energy Contribution 71
      4.5.2.1 Filled Rubbers 71
      4.5.2.2 Block and Graft Copolymers 73
      4.5.2.3 Elastomeric Blends 75
  4.6 Thermomechanics of Bioelastomers 76

5 Thermomechanics of Solid (Glassy and Crystalline) Polymers 76
  5.1 Glassy Polymers 76
    5.1.1 Undrawn Polymers 76
    5.1.2 Drawn Polymers 78
    5.1.3 Microphase Separated Block Copolymers with a Solid Matrix .. 79
  5.2 Crystalline Polymers 80
    5.2.1 Undrawn Polymers 80
    5.2.2 Drawn Polymers 82
      5.2.2.1 Negative Thermal Expansivity of Drawn Cristalline Polymers 82
      5.2.2.2 Thermomechanical Inversion of Internal Energy 85
      5.2.2.3 Thermomechanical Behaviour and Morphology of Drawn
           Crystalline Polymers 87
  5.3 Anisotropy of Thermal Expansivity and Thermomechanical Behaviour . 90
  5.4 Filled Solid Polymers 93
  5.5 Biopolymers 93

6 Concluding Remarks 94

7 List of Symbols and Abbreviations 95

8 References 97

1 Introduction

Thermodynamics of deformation of polymers is a topic which has attracted the attention of scientists for a long time. This is primarily due to the fact that polymers can exhibit a wide variety of behaviour ranging from elasticity of ideal solids to rubberlike elasticity including various inelastic phenomena. Experimental studies of thermomechanical behaviour of rubberlike materials — which has a very long history — have thrown much light on the physical nature of rubberlike elasticity because of the anomalous thermomechanical behaviour of rubbers. Together with the ability to undergo large recoverable deformations it represent the most striking features of rubbers distinguishing them from solids. In particular, it has been proved that rubberlike elasticity is not exclusively entropic and that the intramolecular energy contribution can be very important in deformation of polymer networks and rubberlike materials.

Although the first measurements of a change in temperature of a solid on being stretched or compressed were made by Joule in the 19th century simultaneously with measurements on rubbers, the thermomechanical behaviour of glassy and crystalline polymers has attracted much less attention. As a rule, such experiments are devoted to the tests of the Kelvin relation between stress and temperature. It is well known that this and other similar relations can be obtained in the classical theory of thermoelasticity of solids. However, within the framework of phenomenological thermodynamics it is impossible to give a molecular interpretation of the entropy and internal energy changes without an analysis of a concrete thermomechanical equation of state. The first purpose of this paper is to consider the thermodynamics of deformation of polymers both in solid and rubberlike states with particular emphasis on relationships following from the thermomechanical equations of state for various deformation modes (Sect. 2).

An important step in the thermomechanical study of polymers was taken in the late fifties when deformation calorimetry was first developed by Müller and collaborators. This techniques has now been greatly improved (Sect. 3) and an important progress has been recently achieved owing to the calorimetric behaviour of semicrystalline, glassy and rubbery polymers under deformation. Therefore, the second goal of this paper is to review the information on the thermomechanical behaviour of rubberlike and solid polymers as inferred from deformation calorimetry. We will summarize the current status concerning intra- and interchain effects in non-crystalline and crystalline polymer networks with a special focus on the elucidation of molecular mechanisms of macroscopic deformation and structural changes accompanying the reversible deformation of polymers (Sect. 4 and 5).

The only review which dealt with the polymer thermomechanics both in solid and rubbery state as inferred from deformation calorimetry was published in 1969 [1]. Our paper gives a review of further advances in this field.

2 Thermomechanics of Elastic Systems

2.1 Phenomenological Aspects

If an isotropic condensed system is subjected to several forces of various types, the first law of thermodynamics gives the change in the internal energy U as [2,3)]

$$dU = dQ + \sum_{i=1}^{n} \xi_i \, dx_i \tag{1}$$

where dQ is the element of heat adsorbed, ξ_i is the generalized force conjugated with the generalized coordinate x_i and n is the number of generalized forces. Here, we will consider only the cases when a system is under the action of only two generalized forces, one of which is the external pressure P, since experimental thermomechanical studies of polymers are usually carried out under such conditions. With such assumptions, for a reversible quasistatic process Eq. (1) yields

$$dU = TdS - PdV + \xi \, dx \tag{2}$$

where V is the volume of the system, T the temperature and S the entropy. Using Eq. (2) and the standard definition of the associated free energy F, the enthalpy H and the free enthalpy G, we have

$$dF = -SdT - PdV + \xi \, dx \tag{3}$$

$$dH = TdS + VdP + \xi \, dx \tag{4}$$

$$dG = -SdT + VdP + \xi \, dx \tag{5}$$

The thermodynamic potentials introduced differ from those in use in the thermodynamics of gases and fluids in the additional $\xi \, dx$ terms. For solids, it is convenient to introduce four additional thermodynamic potentials [4-7)]

$$U^* = U - \xi x; \quad dU^* = TdS - PdV - x \, d\xi \tag{6}$$

$$F^* = U^* - TS; \quad dF^* = -SdT - PdV - x \, d\xi \tag{7}$$

$$H^* = U^* - \xi x; \quad dH^* = TdS + VdP - x \, d\xi \tag{8}$$

$$G^* = H^* - \xi x; \quad dG^* = -SdT + VdP - x \, d\xi \tag{9}$$

Using eight thermodynamic potentials introduced, 24 Maxwell relations containing certain partial derivatives can be obtained easily. These relations together with the corresponding specific heats $C_{y,z} = T(dS/dT)_{y,z}$ (where y represents either V or P, and z represents either ξ or x) permit to describe phenomenological relationships between the deformation (or stress) in solids and the accompanying thermal effects.

In particular, the thermal effect of isobaric-isothermal uniaxial extension of a solid rod by the force f can be derived from Eq. (9)

$$(\partial Q/\partial f)_{P,T} = T(\partial S/\partial f)_{P,T} = T(\partial L/\partial T)_{P,f} = T\beta_{P,f}L \tag{10}$$

or in the integral form

$$Q_{P,T} = T\beta_{P,f}Lf \tag{11}$$

where L is the length of the rod and $\beta_{P,f} = 1/L(\partial L/\partial T)_{P,f}$ is the linear thermal expansion coefficient at constant P and f. According to Eq. (11) the extension of a rod with a positive $\beta_{P,f}$ is accompanied by absorption of heat. Such a behaviour is very characteristic of the majority of solids.

If the force is applied instanteneously, the deformation of the bar is adiabatic and is accompanied by a relative temperature change. The temperature change can be derived from Eq. (8):

$$(\partial T/\partial f)_{P,S} = -(\partial L/\partial S)_{P,f} = -\beta_{P,f}LT/C_{P,f} \tag{12}$$

or in the integrated form

$$T_{P,S} = -\beta_{P,f}LTf/C_{P,f} . \tag{13}$$

Equation (13) shows that solids with a positive $\beta_{P,f}$ get cooler at adiabatic extension and warmer under adiabatic compression.

A very important problem in the thermodynamics of deformation of condensed systems is the relationship between heat and work. From Eqs. (2) and (4) by integration, the internal energy and enthalpy can be derived. As in other condensed systems, the enthalpy differs from the internal energy at atmospheric pressure only negligibly, since the internal pressure in condensed systems $P_i \gg P$. Therefore, the work against the atmospheric pressure can be neglected in comparison with the term $\xi_i x_i$. Hence it follows that

$$\Delta U \approx H = T\Delta S + W \tag{14}$$

where $W = \int f_i \, d\xi$.

Using Eq. (14), one can arrive at a convenient form for characterizing the elastic systems by introducing the characteristic ratio [7-9]

$$\eta = Q/W = \Delta U/W - 1 = \omega - 1 \tag{15}$$

Two idealized limiting models can be discussed:
1. The system with ideal elastic energy ($\eta = 0$, $\omega = 1$), in which the reversible work W is totally transformed into the internal energy ΔU. Deformation of such systems is not accompanied by thermal effects. The classical theory of elasticity treats deformation of elastic systems from this point of view.

2. The ideal entropy-elastic system ($\eta = -1$, $\omega = 0$), in which the work exchanged is totally transformed into a change in entropy. A well-known example of such systems is the ideal gas.

Deformation of real condensed bodies can be accompanied by both the internal energy and entropy (or temperature) change, but phenomenological thermodynamics can not answer the question concerning the molecular nature of the changes. This is the consequence of the fact that the phenomenological thermodynamics does not allow to determine the change of the potential energy of the particles of a condensed system under deformation and, therefore, does not permit to determine the change of its temperature or entropy. The answer can be obtained only by analysis of appropriate thermomechanical equations of state. Let us therefore consider a concrete elastic system characterized by the corresponding thermomechanical equation of state. We see that the characteristic ratios η and ω introduced above allow us to describe elastic systems in a particularly convenient way with reference to one of the limiting models.

2.2 Thermomechanics of Quasi-Isotropic Hookean Solids

2.2.1 Uniform (Volume) Dilatation and Compression

The thermomechanical equation of state of such a solid body may be obtained by combining the Hook's law and the law of thermal expansion [7,10]. Hence, for uniform extension we have (as a first approximation)

$$\sigma = K\left[\frac{V}{V_0}(1 - \alpha T) - 1\right] \tag{16}$$

where σ is the stress, K the modulus, V_0 the volume in the unstrained state (at initial temperature T_0), V the volume in the strained state and α the volume thermal expansion coefficient. Equation (16) neglects the variation of K and α with temperature and also the second order terms, such as αT^2. The transition from the extension to compression in Eq. (16) may be carried out by the replacement of σ by $-P$.

The analysis of Eq. (16) has led to the conclusion [7] that the strain-energy function W in the mode of uniform deformation is parabolic with a minimum potential energy in the unstrained state

$$W = Ke^2/2 \tag{17}$$

where $e = (V - V')/V'$, in which V' is the volume in the unstrained state. On the other hand, the heat of deformation is the linear function of deformation

$$Q = T \Delta S = \alpha T\sigma = \alpha TKe \tag{18}$$

The characteristic parameter η of the Hookean solid can be found immediately

$$\eta = 2\alpha T/e. \tag{19}$$

Hence, we arrive at the conclusion that only in the limit $\alpha \to 0$ the Hookean body is the ideal energy-elastic one ($\eta = 0$) and the uniform deformation of a real system is accompanied by thermal effects. Equation (19) shows also that the dependence of the parameter η (as well as ω) on strain is a hyperbolic one and α, the phenomenological coefficient of thermal expansion in the unstrained state, is determined solely by the heat to work and the internal energy to work ratios. From Eqs. (17) and (18), we derive the internal energy of Hookean body

$$\Delta U = Ke^2/2 + \alpha TKe \qquad (20)$$

It can be easily demonstrated that for a Hookean body a thermomechanical inversion of the internal energy ($\Delta U = 0$) must occur at the deformation

$$e_{inv} = -2\alpha T . \qquad (21)$$

We see from Eq. (21) that the internal energy inversion occurs at compression of the system with a positive thermal expansivity and at extension with the negative one. Occurrence of the thermomechanical internal energy inversion in Hookean solids is a result of a different dependence of the work and heat on strain (Fig. 1).

2.2.2 Simple Elongation and Compression

The thermomechanical equation of state of an isotropic Hookean rod subjected to force f along the rod axis can be obtained analogously to that used for the uniform deformation [3,7,8)]

$$\sigma = f/A_0 = E\left[\frac{L}{L_0}(1 - \beta_{P,f}T) - 1\right] \qquad (22)$$

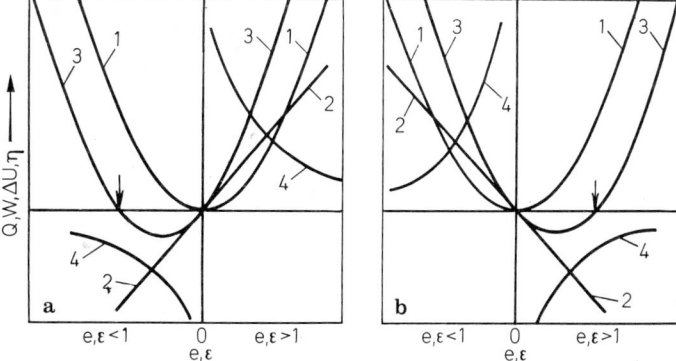

Fig. 1a and b. Mechanical work W (1), elastic heat Q (2), internal energy change ΔU (3) and heat to work ratio η (4) as a function of strain e (uniform deformation) or ε (unidirectional deformation) for quasi-isotropic Hookean solid [8)]. a — positive α and β; b — negative α and β. The arrows indicate inversion points (see text)

where E is Young's modulus, A_0 the cross-sectional area and $\beta_{P,f}$ the linear thermal expansion coefficient under constant P and f.

$$\varepsilon = \frac{L - L'}{L'} \quad \text{with} \quad L' = L_0(1 + \beta_{P,f}T)$$

For most solids, one can neglect the difference between $\beta_{P,f}$ ($\alpha_{P,f}/3$ for an isotropic body) and the coefficient of thermal expansion at constant P is usually used. Therefore, we may use β and α without subscripts. Assuming that E and β are independent of temperature and ignoring the change in lateral dimensions during deformation (i.e. we take the Poisson's ratio $\mu = 0$, because this simplification gives effects of only the second order of smallness), one can arrive at relations similar to Eqs. (17)–(21). To do this, it is necessary to replace in Eq. (16) the volume deformation e by ε, the modulus K by E and α by β (see Fig. 1). For the simple deformation of a Hookean body the characteristic parameter η is also inversely dependent on strain, viz. $\eta = 2\beta T/\varepsilon$ and $\varepsilon_{inv} = -2\beta T$. It is interesting to note that

$$\varepsilon_{min} = -\beta T \tag{23}$$

$$\Delta U_{min} = -E(\beta T)^2/2 \tag{24}$$

Since the values of β and α for solid polymers are usually in the range of $10^{-4} - 10^{-5}$ K^{-1}, the thermomechanical inversion of the internal energy must occur at the deformation of a few percent [7].

2.2.3 Thermal Expansivity

The above analysis has demonstrated that thermal effects resulting from the reversible uniform and uniaxial deformations are determined by the coefficients of thermal expansion. The chain structure of polymers leads to a large local anisotropy of the thermal expansivity because of a large difference in inter- and intrachain interaction potentials. Moreover, from the concept of the contribution of vibrational modes in chain structures to the heat capacity one may state that the thermal expansivity along the polymer chains must be close to zero below and at room temperature, because at this temperature only the deformational and torsional modes can be completely exited [7,11]. However, the theoretical consideration of thermal properties of chain and planar structures show that such structures can exhibit a negative thermal expansion along the chain or in plane [12], which is a consequence of an important role of bending waves with an unusual dispersion relationship in such structures ($\omega_t \sim k^2$ instead of $\omega_l \sim k$, where ω is the frequence and k is the wave number). The recent model theoretical analysis of the negative thermal expansion of polymer crystals has also supported the existence of a negative thermal expansion coefficient along the polymer chains [13-15]. In particular, the calculations carried out by Choy et al. [14,15] have led to $\beta_{cr\parallel} = -1.3 \times 10^{-5}$ K^{-1}.

Negative thermal expansion coefficients along the chain axes have been observed experimentally on many crystalline polymer lattices [7,16-18]. Hence, the thermal contraction along the chain axis seems to be a general phenomenon in crystalline polymers. As a result of this conclusion, we are immediately faced with the question what the consequences are for crystalline polymers of the negative thermal expansivity along the c-axis of the crystalline lattices. For bulk unoriented polymers, it seems to have no effect because of the following reasons. First, the absolute value of the negative thermal expansion coefficient is much smaller than that of two other positive coefficients perpendicular to the chain axis, which leads to the positive volume thermal expansivity. Because of statistical arrangement of crystallites in unoriented crystalline polymers and, consequently, the chain axes, it is clear that the thermal expansivity in macroscopic samples along any direction must be a positive one. The second reason is the presence of amorphous regions in crystalline polymers. In contrast to crystalline lattices, the anisotropy of the thermal expansivity in the amorphous regions seems to be absent. These arguments allow to suppose that the thermal shrinkage of crystalline macromolecules will have no consequence for macroscopic thermal properties of unoriented crystalline polymers. However, the situation may change in oriented crystalline polymers due to orientation of both crystallites and amorphous regions. Hence, in oriented crystalline polymers one can expect to observe an anomalous thermomechanical behaviour.

2.2.4 Shear. Torsion

According to the theory of elasticity, the pure shear and torsion are accompanied only by a change in shape of an elastic body but its volume remains unchanged. The equation of state of the Hookean body for pure shear can be represented in the following form [7,10]

$$\tau = G_0(1 - mT)\gamma_s \qquad (25)$$

where τ is the shear stress, γ_s the shear deformation, G_0 the shear modulus and m a small positive constant. Equation (25) yields the following expressions for work W and heat Q

$$W = \tau\gamma_s/2 = G_0(1 - mT)\gamma_s^2/2 \qquad (26)$$

$$Q = TG_0 m\gamma_s^2/2. \qquad (27)$$

Comparing W and Q, we find $\eta = mT/(1 - mT) \ll 1$. Hence, we see that the heat to work ratio for shear is independent of deformation. It means that thermal effects resulting from pure shear can be ignored (as a first approximation). This is also true for torsion.

This result stresses the fact that the heat effects can be detected only at such deformation modes of quasi-isotropic Hookean solids that are accompanied by a change in volume. Thus, the thermal effects are the consequence of the change of the vibrational entropy which in turn is a result of the volume change. It is very important to emphasize now that the internal energy and entropy changes are closely interrelated and their values are of the same order of magnitude.

2.3 Thermomechanics of Molecular Networks

In the above thermomechanical equations of state we neglect the variation of K, E, α and β with temperature and strain (or stress). As we will see later, this is a good first approximation for the majority of solids and liquids including polymers. However, there is a class of elastic bodies, e.g. molecular networks and rubberlike materials, for which this assumption is not fulfilled in principle for the majority of the deformation modes. On simple elongation, rubberlike systems are capable to undergo very large reversible elastic deformations. The modulus of elasticity, which in this case is strongly dependent on deformation, is some order of magnitude lower than the bulk modulus. Unlike the solids, the modulus of elasticity E of the deformed networks and rubberlike materials is proportional to the absolute temperature (excluding very initial deformations). A very striking feature of the thermomechanical behaviour of elastomers is a strong dependence of the linear thermal expansivity on deformation. This dependence is as follows: the initial positive thermal expansivity decreaes with deformation and in the vicinity of 8–10% deformation the expansivity becomes negative and at moderate deformations it reaches the value of 10^{-3} K^{-1} typical for gases. All these facts demonstrate that the thermomechanical behaviour of rubberlike bodies differs in principle from that of solids. It is now very important to emphasize that high elastic deformations are characteristic only for those deformation modes which are connected with the elasticity of the form. Since the volume compressibility of rubberlike bodies is very small (the same order as for liquids), the thermodynamics of their uniform (volume) deformation is described by Eqs. (17)–(21).

2.3.1 Thermomechanical Equations of State Based on Statistical Theories

2.3.1.1 Thermomechanics of Gaussian Networks

2.3.1.1.1 Uniaxial Deformation

The classical statistical theory of rubber elasticity[1]) for a Gaussian polymer network which took into account not only the change of conformational entropy of elastically active chains in the network but also the change of the conformation energy, led to the following equation of state for simple elongation or compression [19–21)

$$f = vkTL_0^{-1}(\langle r^2 \rangle / \langle r^2 \rangle_0)\left(\lambda - \frac{V}{V_0 \lambda^2}\right) \tag{28}$$

or alternatively

$$f = vkTL_i^{-1}(\langle r^2 \rangle_i / \langle r^2 \rangle_0)(\alpha_* - \alpha_*^{-2}) \tag{29}$$

where L_0 and V_0 are length and volume of the network at zero force, zero pressure and temperature T; L and V are the corresponding quantities at force f, pressure P and temperature T; $\lambda = L/L_0$ is the elongation (or compression), v the number of elastically active chains in the network, k the Boltzmann constant, $\langle r^2 \rangle$ the mean square end-to-end distance of the network chains in volume V_0, $\langle r^2 \rangle_0$ that of the corresponding free chains, $\langle r^2 \rangle_i$ the mean square end-to-end distance of a network in the undistorted state of volume V and $\alpha_* = L/L_i$, L_i the length of the undistorted

---

1 The newer development of the theory is analysed in Sect. 2.3.3

sample at volume V. The value of α_* differs negligibly from λ. Below we will use the equation of state only in the form of Eq. (28).

The molecular theory of elasticity of polymeric networks which leads to the equation of state, Eq. (28), rests on the following basic postulates: Undeformed polymeric chains of elastic networks adopt random configurations or spatial arrangements in the bulk amorphous state. The stress resulting from the deformation of such networks originates within the elastically active chains and not from interactions between them. It means that the stress exhibited by a strained network is assumed to be entirely intramolecular in origin and intermolecular interactions play no role in deformations (at constant volume and composition).

Making use the equation of state in Eq. (28), the work, heat and internal energy change at V, T = const. can be derived

$$W_{V,T} = \frac{C}{2} \frac{(\lambda - 1)}{\lambda} (\lambda^2 + \lambda - 2) \tag{30}$$

$$Q_{V,T} = T(\Delta S)_{V,T} = -\frac{C}{2}\left(1 - T\frac{d \ln \langle r^2 \rangle_0}{dT}\right) \frac{(\lambda - 1)}{\lambda} (\lambda^2 + \lambda - 2) \tag{31}$$

$$\Delta U_{V,T} = \frac{C}{2} T \frac{d \ln \langle r^2 \rangle_0}{dT} \frac{(\lambda - 1)}{\lambda} (\lambda^2 + \lambda - 2). \tag{32}$$

In these equations $C = vkTL_0^{-1} \langle r^2 \rangle / \langle r^2 \rangle_0$.

The characteristic parameters

$$\eta_{V,T} = (Q/W)_{V,T} = -1 + T\frac{d \ln \langle r^2 \rangle_0}{dT} = (f_s/f) \tag{33}$$

$$\omega_{V,T} = (\Delta U/W)_{V,T} = T\frac{d \ln \langle r^2 \rangle_0}{dT} = (f_u/f) \tag{34}$$

are usually called the entropic and energetic components of the work (force). According to the Gaussian theory of rubber elasticity they should be constant and independent of deformation.

Equations (30)–(34) demonstrate very obviously the basic idea of the statistical theory of rubber elasticity. On deformation of the chains, some part of the work is connected with the intramolecular energy change resulting from the transition of the chains from one spatial configuration to another. The sign and value of the energy change, are dependent on the chemical structure of the macromolecules. The parameter $d \ln \langle r^2 \rangle_0/dT$ may be both positive and negative [20] and, consequently, the simple deformation of polymer networks may be accompanied by both increasing and decreasing of the internal energy. For the chains with free rotational states, $d \ln \langle r^2 \rangle_0/dT = 0$, $\eta = -1$ which corresponds to the ideal entropy-elastic model. Hence, the factor $d \ln \langle r^2 \rangle_0/dT$ may be considered as a parameter of non-ideality of polymer chains [21]. Unlike gases, which can behave ideally by decreasing the pressure, this type of non-ideality can not be removed by diluting since it is intramolecular in origin.

Although Eqs. (31) and (32) are very simple, it is necessary to carry out the measurements at constant volume. This experiment is difficult in practice. Deformation of rubbers is usually accompanied by a change in volume $\Delta V/V \approx 10^{-4}$. This volume change can be neglected for determining the mechanical work. However, even small change in volume during deformation are extremely important for determining the vibrational entropy and the internal energy. Thus, even a small change of the volume can strongly distort the intramolecular effects.

Achievement of the constant volume condition requires applying hydrostatic pressure and there is only one reliable set of experiments [22, 23] which will be discussed later. Usually the thermomechanical experiments are carried out at the constant pressure condition. The expression for W, Q, ΔU, η and ω under P, T = const. are

$$W_{P,T} = \frac{C}{2} \frac{(\lambda - 1)}{\lambda} (\lambda^2 + \lambda - 2) \tag{35}$$

$$Q_{P,T} = -\frac{C}{2} \left[\left(1 - T \frac{d \ln \langle r^2 \rangle_0}{dT}\right) - \frac{2\alpha T}{\lambda^2 + \lambda - 2} \right] \frac{(\lambda - 1)}{\lambda} (\lambda^2 + \lambda - 2) \tag{36}$$

$$U_{P,T} = \frac{C}{2} \left(T \frac{d \ln \langle r^2 \rangle_0}{dT} + \frac{2\alpha T}{\lambda^2 + \lambda - 2} \right) \frac{(\lambda - 1)}{\lambda} (\lambda^2 + \lambda - 2) \tag{37}$$

$$\eta = \left(\frac{Q}{W}\right)_{P,T} = -1 + T \frac{d \ln \langle r^2 \rangle_0}{dT} + \frac{2\alpha T}{\lambda^2 + \lambda - 2} \tag{38}$$

$$\omega = \left(\frac{U}{W}\right)_{P,T} = T \frac{d \ln \langle r^2 \rangle_0}{dT} + \frac{2\alpha T}{\lambda^2 + \lambda - 2} \tag{39}$$

These expressions demonstrate that the change of entropy and internal energy on deformation under these conditions is both intra- and intermolecular in origin. Intramolecular (conformational) changes, which are independent of deformation, are characterized by the temperature coefficient of the unperturbed dimensions of chains $d \ln \langle r^2 \rangle_0/dT$. The intermolecular changes are characterized by the thermal expansivity α and are strongly dependent on deformation. The difference between the thermodynamic values under P, T = const. and V, T = const. is very important at small deformations since at $\lambda \to 1$ $2\alpha T/(\lambda^2 + \lambda - 2)$ tends to infinity.

Comparing Eqs. (33), (34) and (38), (39), we arrive immediately at

$$\lambda_{\Delta V} = \eta_{P,T} - \eta_{V,T} = \frac{2\alpha T}{\lambda^2 + \lambda - 2} \tag{40}$$

$$\omega_{\Delta V} = \omega_{P,T} - \omega_{V,T} = \frac{2\alpha T}{\lambda^2 + \lambda - 2}. \tag{41}$$

Thermomechanics of Polymers

These expressions characterize the relative change of the entropy and internal energy resulting from the volume change. Therefore, they must be identical with the corresponding expressions for solids. It can be easily proved by introducing the strain ε instead of λ into Eq. (40) and (41) and neglecting ε^2 terms

$$\left(\frac{Q}{W}\right)_{\Delta V} = \left(\frac{\Delta U}{W}\right)_{\Delta V} = \frac{2\alpha T}{3\varepsilon} = \frac{2\beta T}{\varepsilon}. \tag{42}$$

This expression is fully identical with the expression for η for simple elongation of solids (see 2.2.2).

The absolute values of the intermolecular change of the internal energy and entropy associated with the volume dilation may be written as

$$T(\Delta S)_{\Delta V} = (\Delta U)_\Delta = C\alpha T \frac{\lambda - 1}{\lambda}. \tag{43}$$

We recognize from Eq. (43) that the change of internal energy resulting from the volume change of a Gaussian network is exactly balanced by the equivalent change of entropy and, thus, this volume change gives no contribution to the free energy of deformation [24, 25]. Provided α and the isothermal compressibility $\varkappa$ are independent of strain, by using the known thermodynamic equality $(dU/dV)_{P,T} = \alpha KT$, one can obtain the equation for the volume change

$$\frac{\Delta V}{V_0} = \frac{C}{K} \frac{(\lambda - 1)}{\lambda} = C\varkappa \frac{(\lambda - 1)}{\lambda}. \tag{44}$$

All these expressions demonstrate the distinction between the mechanism of elasticity of solids and elastomers on simple deformation (at the condition of constant pressure and temperature). The elasticity of a solid is a result of the resistance to change of its volume, and all the thermodynamic properties are connected with the volume change. The elasticity of an elastomeric body is a result of the resistance to change of its shape and the work is spent only on the conformational change of chains. A small volume change accompanying the change in shape is an attendent effect, since the resulting energy change is fully compensated by the entropy change.

Equations (36) and (37) predict the inversions of heat and internal energy on deformation [24]. The inversion of heat must occur at (Fig. 2)

$$\lambda_Q = 1 + \frac{2}{3} \frac{\alpha T}{(1 - T\, d \ln \langle r^2 \rangle_0 / dT)}. \tag{45}$$

This inversion of heat is due to a competition between the increase of the vibrational entropy connected with the volume change at deformation and the decrease of the conformational entropy. The deformation at which a maximum of heat is absorbed at elongation is given by

$$\lambda_Q^{max} = 1 + \frac{1}{3} \frac{\alpha T}{(1 - T\, d \ln \langle r^2 \rangle_0 / dT)} \tag{46}$$

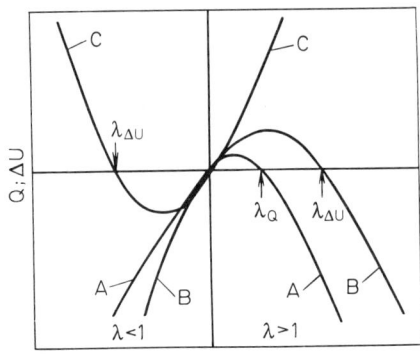

Fig. 2. Heat and internal energy changes as a function of deformation at P, T = const. for a Gaussian polymer networl [7,8]. A — heat; B — internal energy (d ln $\langle r^2 \rangle_0/dT < 0$); C — internal energy (d ln $\langle r^2 \rangle_0/dT > 0$). The arrows indicate inversion points (see text)

and Q^{max} at this deformation is given by

$$Q^{max} \approx \frac{C}{3} \frac{(\alpha T)^2}{(1 - T\, d \ln \langle r^2 \rangle_0/dT)}. \tag{47}$$

For the internal energy inversion point, we arrive at

$$\lambda_U = 1 - \frac{2}{3} \frac{\alpha T}{T\, d \ln \langle r^2 \rangle_0/dT}. \tag{48}$$

Unlike the thermomechanical inversion of heat, the inversion of internal energy is possible only for chains with d ln $\langle r^2 \rangle_0/dT \neq 0$. It is also evident from Eq. (48) that for d ln $\langle r^2 \rangle_0/dT < 0$ the inversion must occur at $\lambda > 1$ and vice versa. For values $\alpha = (6 - 10) \times 10^{-4}$ K^{-1} and d ln $\langle r^2 \rangle_0/dT = (5 - 0.5) \times 10^{-3}$ K^{-1}, which are typical for polymeric networks, $\lambda_U = 1.3 - 2.2$ for extension and $\lambda_U > 0.5$ for compression. It must be emphasized that this thermomechanical inversion of internal energy is not connected with the stress-induced crystallization and arises from the different signs of inter- and intrachain contributions to the internal energy. The extreme of the internal energy occurs at the deformation

$$\lambda_U^{ex} = 1 - \frac{1}{3} \frac{\alpha T}{T\, d \ln \langle r^2 \rangle_0/dT}. \tag{49}$$

It is quite obvious that the thermomechanical inversions both of heat and internal energy must disappear at the condition of constant volume.

Finally, making use of the equation of state of Eq. (28) one can demonstrate that the linear thermal expansion coefficient depends on the deformation as follows (for isoenergetic chains) [26,27]

$$\beta_\parallel(\lambda) = -\frac{1}{T} \frac{\lambda^3 - 1}{\lambda^3 + 2} + \frac{\alpha}{\lambda^3 + 2} \tag{50}$$

$$\beta_\perp(\lambda) = \frac{1}{2T} \frac{\lambda^3 - 1}{\lambda^3 + 2} + \frac{\alpha}{2} \frac{\lambda^3 + 1}{\lambda^3 + 2}. \tag{51}$$

In the undeformed state, $\beta_\parallel = \beta_\perp = \alpha/3 > 0$. At large λ, one can neglect the terms with α and we arrive at

$$\beta_\parallel \approx -\frac{1}{T}\frac{\lambda^3 - 1}{\lambda^3 + 2} \tag{52}$$

$$\beta_\perp \approx \frac{1}{2T}\frac{\lambda^3 - 1}{\lambda^3 + 2}. \tag{53}$$

These expressions show that a deformed polymer network is an extremely anisotropic body and possesses a negative thermal expansivity along the orientation axis of the order of the thermal expansivity of gases, about two orders higher than that of macromolecules incorporated in a crystalline lattice (see 2.2.3). In spite of the large anisotropy of the linear thermal expansivity, the volume coefficient of thermal expansion of a deformed network is the same as of the undeformed one. As one can see from Eqs. (50) and (51) $\beta_\parallel + 2\beta_\perp = \alpha$. Equation (50) shows also that the thermoelastic inversion of $\beta_\parallel$ must occur at λ_{inv} $(\varepsilon_{inv}) \approx 1 + (1/3)\alpha T$. It coincides with λ_Q^{max} for isoenergetic chains [see Eq. (46)].

In the current statistical theory of rubber elasticity, it is suggested that the front-factor $\langle r^2 \rangle / \langle r^2 \rangle_0$ is independent of volume (and hence of pressure). Tobolsky and Shen [28, 29], however, have supposed that such a dependence occurs due to intermolecular forces. They have proposed a semiempirical equation of state taking into account the dependence

$$f = \frac{\nu kT}{L_0} \frac{\langle r^2 \rangle}{\langle r^2 \rangle_0} \left(\frac{V_0}{V}\right)^\gamma \left(\lambda - \frac{V}{V_0 \lambda^2}\right) \tag{54}$$

where $\gamma = (\partial \ln \langle r^2 \rangle_0 / \partial \ln V)_{T,L}$ is a constant dependent on the chemical structure of chains. If γ is taken to be zero, Eq. (54) is transformed into Eq. (28). From the equation of state, the energetic contribution of the work is readily derived as

$$\left(\frac{\Delta U}{W}\right)_{V,T} = \left(\frac{\Delta U}{W}\right)_{P,T} - \frac{2\alpha T}{\lambda^2 + \lambda - 2} - \gamma \alpha T. \tag{55}$$

Equation (55) differs from Eq. (39) by the term $(-\gamma \alpha T)$ and hence $(\Delta U/W)_{V,T}$ should be independent of extension ratio λ, since $(-\gamma \alpha T)$ is not a function of deformation. Equation (54) leads also to the following expression for the strain-induced volume dilation

$$\frac{\Delta V}{V_0} = C\varkappa \frac{(\lambda - 1)}{\lambda}\left[1 + \frac{\gamma}{2}(\lambda^2 + \lambda - 2)\right]. \tag{56}$$

If γ is taken to be zero, Eq. (56) is identical with Eq. (44). The change of entropy and internal energy corresponding to the volume dilation according to Eq. (56) are

$$(\Delta U)_{\Delta V} = T(\Delta S)_{\Delta V} = C\alpha T \frac{(\lambda - 1)}{\lambda}\left[1 + \frac{\gamma}{2}(\lambda^2 + \lambda - 2)\right]. \tag{57}$$

Recently, an attempt has been made to give a molecular interpretation of the parameter γ [30].

Schwarz [31] has modified the equation of state for the Gaussian polymer network by introducing a coefficient taking into account the influence of the packing of chains and their sizes on the mean square end-to-end distance of deformed chains.

2.3.1.1.2 Torsion

Flory [19] has shown that for Gaussian networks the connection between the temperature dependence of the force at constant volume and the temperature coefficient of molecular dimensions $d \ln \langle r^2 \rangle_0/dT$ holds for all types of distortions. According to Treloar [32], the equation of state for torsion of the Gaussian networks is

$$M = \left(\frac{\pi\nu kT}{2V_0}\right)\left(\frac{\langle r^2 \rangle}{\langle r^2 \rangle_0}\right)\left(\frac{\theta}{L_0}\right) a_0^4 \tag{58}$$

where M is the torque which is needed for twisting a cylinder of length L_0 by the angle θ; a_0 is the radius of an unstrained sample. From Eq. (58) we get

$$W = \int_0^\theta M\, d\theta \left(\frac{1}{4}\frac{\pi\nu kT}{V_0}\right)\left(\frac{\langle r^2 \rangle}{\langle r^2 \rangle_0}\right) \cdot \frac{a_0^4}{L_0} \theta^2 \tag{59}$$

$$T(\Delta S)_{P,T,L} = W(-1 - \alpha T + T\, d \ln \langle r^2 \rangle_0/dT) \tag{60}$$

$$(\Delta U)_{P,T,L} = W(-\alpha T + T\, d \ln \langle r^2 \rangle_0/dT) \tag{61}$$

$$T(\Delta S/W)_{P,T,L} = -1 - \alpha T + T\, d \ln \langle r^2 \rangle_0/dT \tag{62}$$

$$(\Delta U/W)_{P,T,L} = -\alpha T + T\, d \ln \langle r^2 \rangle_0/dT \tag{63}$$

$$T(\Delta S)_{V,T,L} = W(1 - T\, d \ln \langle r^2 \rangle_0/dT) \tag{64}$$

$$(\Delta U)_{V,T,L} = WT\, d \ln \langle r^2 \rangle_0/dT \tag{65}$$

$$T(\Delta S/W)_{V,T,L} = -1 + T\, d \ln \langle r^2 \rangle_0/dT \tag{66}$$

$$(\Delta U/W)_{V,T,L} = T\, d \ln \langle r^2 \rangle_0/dT \tag{67}$$

We see that the relative change of entropy and internal energy at constant pressure is independent of the degree of twisting. This conclusion differs from that obtained for simple extension or compression. The entropic and energetic components in torsion are identical with the result for simple deformation. Equations (60) and (61) lead to the conclusion that there can be no thermomechanical inversions of heat and internal energy in torsion.

The change of entropy and internal energy, associated with the change of volume due to torsion is given by the equation

$$T(\Delta S)_{\Delta V} = (\Delta U)_{\Delta V} = -W\alpha T . \tag{68}$$

The corresponding change of volume is

$$\Delta V = -\frac{M\varkappa\theta}{2} = -\left(\frac{1}{4}\frac{\eta \cdot v \cdot k \cdot T}{V_0}\right)\left(\frac{\langle r^2 \rangle}{\langle r^2 \rangle_0}\right)\frac{\theta^2}{L_0}a_0^4\varkappa \tag{69}$$

Hence, from Eqs. (68) and (69) one can conclude that the change in volume due to torsion does not contribute to the free energy of deformation, that torsion is accompanied by a decrease in volume and that this volume decrease is a parabolic function of the twisting angle.

2.3.1.2 Thermomechanics of Non-Gaussian Networks

Equations (28) and (29) are derived from the statistical theory based on the Gaussian statistics which describes the network behaviour if the network is not deformed beyond the limit of the applicability of the Gaussian approximation [33]. For long chains, this limit is close to 30% of the maximum chain extension. For values of r, which are comparable with r_{max}, the force-strain dependence is usually expressed using the inverse Langevin function [33, 34]

$$f = \frac{vkT}{3}p^{1/2}[\mathscr{L}^{-1}(\lambda p^{-1/2}) - \lambda^{-3/2}\mathscr{L}^{-1}(\lambda^{-1/2}p^{-1/2})] \tag{70}$$

where $\mathscr{L}^{-1}$ is the inverse Langevin function, p is the number of statistical segments in a chain, which determines the limit chain extensibility $\lambda = p^{1/2}$. If $\lambda_{max} = \sqrt{p}$, the force f becomes infinite since $\mathscr{L}^{-1}(1) \to \infty$. According to estimations of Treloar [33], the difference in the values of force f corresponding to Eqs. (28) and (70) is about 3% at $r = 0.3 r_{max}$ and about 8% at $r = 0.5 r_{max}$.

To elucidate the role of the limited extensibility of chains on the thermomechanical properties of non-Gaussian networks, Schwarz [35-37] carried out theoretical considerations of the behaviour of short-chain model networks consisting of chains with 4-14 bonds. As a result of calculations assuming no fluctuations of crosslinks, it was concluded that the energetic component resulting from simple elongation is equal to the theoretical value. It has been demonstrated that the energy contribution depends on deformation, being a very weak linear function of the extension for the networks with positive energetic contribution and a strong one for the networks with a negative energetic component. This is a rather unexpected conclusion and we think that this problem needs further theoretical consideration.

In other statistical theories of rubber elasticity (see e.g. reviews [29, 34]) the Gaussian statistics is not valid even at small deformations and the intramolecular energy component is dependent on deformation.

The non-Gaussian theories of rubber elasticity have the disadvantage of containing parameters which generally can be determined only by experiment. Recently,

Curro and Mark [38] have proposed a new non-Gaussian theory of rubber elasticity based on rotational isomeric state simulations of network chain configurations. Specifically, Monte Carlo calculations were used to determine the distribution functions for end-to-end dimensions of the network chains. The utilization of these distribution functions instead of the Gaussian function yields a large decreases in the entropy of the network chains.

2.3.2 Phenomenological Equations of State

It is well known that the equation of state of Eq. (28) based on the Gaussian statistics is only partially successful in representing experimental relationships tension-extension and fails to fit the experiments over a wide range of strain modes [29,33,34]. The deviations from the Gaussian network behaviour may have various sources discussed by Dušek and Prins [34]. Therefore, phenomenological equations of state are often used. The most often used phenomenological equation of state for rubber elasticity is the Mooney-Rivlin equation [29,33,34]

$$\frac{f}{A_0} = \left(2C_1 + \frac{2C_2 V_0}{V\lambda}\right)\left(\lambda - \frac{V}{V_0 \lambda^2}\right) \tag{71}$$

where C_1 and C_2 are constants.

Shen [39] has considered the thermoelastic behaviour of the materials described by the Mooney-Rivlin equation and has shown that the energetic component is given by

$$\left(\frac{f_U}{f}\right) = 1 - \frac{\partial \ln f}{\partial T} - \alpha T \left[\frac{1}{(\lambda^3 - 1)} + \frac{1}{\frac{C_1}{C_2}\lambda + 1}\right]. \tag{72}$$

Equation (72) indicates that the energy contribution is a function of deformation and must decrease with an increasing extension ratio. The change in volume resulting from the deformation is given by

$$\Delta V/V_0 = 2\varkappa(\lambda - 1)(C_1/\lambda + C_2). \tag{73}$$

Shen [39] has also considered the thermoelastic behaviour of another widely used phenomenological equation of state, the so-called Valanis-Landel equation. Valanis and Landel [40] have postulated that the stored energy function W should be expressible as the sum of three independent functions of principle extension ratios. This hypothesis leads to the following equation of state

$$f = CA_0 \left[2 \ln \lambda + \left(\frac{V}{V_0 \lambda^3}\right)^{1/2} \ln\left(\frac{V}{V_0 \lambda}\right)\right]. \tag{74}$$

On the basis of Eq. (74), one can get the expression for the energetic contribution

$$\frac{f_u}{f} = \left(1 - \frac{\partial \ln C}{\partial \ln T} - \frac{\alpha T}{3}\right) - \frac{\alpha T}{3}\left[1 - \frac{2\lambda^{3/2} + 2}{(2\lambda^{3/2} + 1)\ln \lambda}\right]. \tag{75}$$

Again, we see that the energy contribution decreases with increasing λ.

It is very important to stress that the decrease of the internal energy contribution with increasing extension ratio is due to a decrease of the intermolecular interaction with deformation, since the intramolecular contribution is independent of the deformation in full accord with the statistical theory. At very high strains, the λ-dependent part of f_u/f approaches a limiting value of -0.68 for the Mooney-Rivlin and -0.07 for the Valanis-Landel materials.

A comprehensive consideration of new phenomenological equations of state for rubber elasticity have been carried out lately by Tschoegl et al.[41-45]. One of their equations of state is given by

$$f = \frac{2C}{n}\left(\frac{V}{V_0}\right)^\gamma \left(\lambda^{n-1} - \frac{(V/V_0)^{n/2}}{\lambda^{(n+2)/2}}\right) \tag{76}$$

where n is the parameter which characterizes the nonlinear stress-strain behaviour. If $n = 2$, this equation reduces into the Tobolsky-Shen equation [Eq. (54)], and if in addition $\gamma = 0$ it transforms into the equation of state of Eq. (28). It was supposed that the constant n is independent of temperature. In this case, the energy contribution corresponding to Eq. (76) should be independent of the deformation ratio. Later, a modification of the theory was proposed which took into account the temperature dependence of the strain parameter n, and this modification led to a dependence of the energy contribution on extension.

Recently, a number of attempts have been made to represent the stored energy function W by a power series containing more than two terms[46-49]. In particular, Chadwick[47] modified the expression for the stored energy function proposed by Ogden[46], and introduced the scalar response function for a compressible elastomer. The generalized form of the stored energy function W (per unit volume) is represented in the form of the following series

$$F = \sum_r (\mu_r/\mu\alpha_r)(\lambda_1^r + \lambda_2^r + \lambda_3^r - 3J^{1/3}\alpha_r) \tag{77}$$

where J is the ratio of the densities of the deformed and undeformed materials and μ_r, α_r and μ are material constants, μ being the shear modulus. We see that this equation is still consistent with the Valanis-Landel hypothesis. Equation (77) leads to the following expression for the volume dilation on extension

$$(\partial \ln V/\partial \lambda)_{P,T} = {}^1/_2\varkappa \Sigma(\alpha_r\mu_r\lambda^{-1/2}\alpha_r - 1) \tag{78}$$

If $\alpha_r = 2$, $\mu_r = 2C_1$, $\alpha_r = -2$ and $\mu_r = -2C_2$, this equation reduces into Eq. (73) for Mooney-Rivlin materials. A further development of this approach is given in [48,49].

In a recent series of papers, Kilian[9,50-52] proposed a new phenomenological approach to rubber elasticity and suggested a molecular network might be considered as a formelastic fluid the conformational abilities of which were adequately characterized by the model of a van der Waals conformational gas with weak interaction. The ideal network is treated as an ideal conformational gas. According to these assumptions, the von der Waals equation of state of real gases yields a thermomechanical equation of state for real networks under simple deformation

$$f = \frac{vkT}{L_0} \frac{\langle r^2 \rangle}{\langle r^2 \rangle_0} D(B - aD) \tag{79}$$

where $D = \lambda - \lambda^2$, $B = \frac{D_m}{D_m - D}$, $D_m = \lambda_m - \lambda_m^2$, $\lambda_m = L_m/L_0$ is the strain related to the limited chain extensibility. The van der Waals parameter, a, takes into account weak interactions between chains. It may be represented by

$$a = a_1(1 - a_2/\lambda^2). \tag{80}$$

Thus, this phenomenological equation includes the three parameters λ_m, a_1 and a_2 which can be obtained from the analysis of the shape of the stress-strain curve.

Starting from Eq. (79), Kilian [9, 50] obtained the following expressions for the entropic and energetic components of the elastic force f in simple extension

$$\left(\frac{f_s}{f}\right)_{P,T} = \left(1 - T\frac{d \ln \langle r^2 \rangle_0}{dT}\right)$$

$$- \beta T \left(\frac{3}{\lambda^3 - 1} + \frac{\bar{B} - a\bar{D} - \bar{a}D + (\partial a/\partial T) D/\beta}{B - aD}\right) \tag{81}$$

$$\left(\frac{f_H}{f}\right)_{P,T} = T\frac{d \ln \langle r^2 \rangle_0}{dT} + \beta T \left(\frac{3}{\lambda^3 - 1} + \frac{\bar{B} - a\bar{D} - \bar{a}D - (\partial a/\partial T) D/\beta}{B - aD}\right) \tag{82}$$

where

$$\bar{B} = \frac{1}{D_m - D}\left(\bar{D}_m - D_m\frac{\bar{D}_m - \bar{D}}{D_m - D}\right); \quad \bar{D}_m = \lambda_m + 2\lambda_m^{-2};$$

$$\bar{D} = \lambda - 2\lambda^{-2};$$

$$\frac{\partial a}{\partial T} = \frac{\partial a_1}{\partial T}\left(1 - \frac{a_2}{\lambda^2}\right) - a_1\lambda^{-2}\frac{\partial a_2}{\partial T}, \quad \bar{a} = 2a_1 a_2 \lambda^{-2}.$$

Integration of these equations leads to $W(\lambda)$, $Q(\lambda)$ and $\Delta U(\lambda)$. Kilian showed that the energetic component $(f_u/f)_{V,T} = (\Delta U/W)_{V,T} = d \ln \langle r^2 \rangle_0/dT$ is in full agreement with the conclusions of statistical theory of rubber elasticity. Hence, if $d \ln \langle r^2 \rangle_0/dT$ and β are known the full energy balance on deformation can be calculated using parameters λ_m, a_1, a_2, and $\partial a/\partial T$.

Making use of the expression

$$(\partial V/\partial L)_{P,T} = (\partial f/\partial P)_{T,L} \tag{83}$$

the change of volume with deformation is expressed by

$$\frac{\Delta V}{V_0} = \varkappa_L L_0^{-2} \int f \left(\frac{3}{\lambda^3 - 1} + \frac{\bar{B} - a\bar{D} - \bar{a}D + \varkappa^{-1}(\partial a_1/\partial P) D}{B - aD} \right) \quad (84)$$

with $\partial a/\partial P = (\partial a_1/\partial P)(1 - a_2/\lambda^2)$ and $\varkappa_L = -(\partial \ln L_0/\partial P)_{T,L}$ being the linear isothermal compressibility in the unstrained state. These volume changes are dependent not only on the compressibility of the network but also on the van der Waals parameters and the pressure coefficient of the interaction parameter a.

The concept of the van der Waals conformational gas yields also a reduced equation of state [52]

$$\tilde{f}' = d \left(\frac{8t}{3d} - 3d \right) \quad (85)$$

where $\tilde{f}' = f/f_c$, $d = D/D_c$, $t = T/T_c$ with the critical coordinates $T_c = 8aD_m/27$, $D_c = D_m/3$ and $f'_c = aD_m^2/27$. The stability of a network under simple elongation is closely controlled by van der Waals parameters. Thus, one can conclude that this new phenomenological equation of state for rubber elasticity is a very promising approach to the thermomechanical behaviour of polymer networks and rubberlike materials.

2.3.3 The New Developments of the Theory of Elasticity of Polymer Networks

During the last decade, the classical theory of rubber elasticity has been reconsidered significantly. It has been demonstrated (see, e.g. Ref. [53]) that, for the phantom non-interacting network whose chains move freely one through the other, the equations of state of Eqs. (28) and (29) for simple deformation as well as for W, Q and ΔU [Eqs. (30)–(32) and (35)–(37)] are proportional not to ν but to ξ, which is the cycle rank of the network, i.e. the number of independent circuits it contains. For a perfect phantom network of uniform functionality $\varphi(>2)$

$$\xi = \nu(\varphi - 2)/\varphi . \quad (86)$$

The behaviour of real networks varies between two extremes [53]: the phantom network model and the affine network model. A number of theoretical models have been formulated recently to describe the behaviour of real networks with sterical restrictions resulting from interchain constraints of fluctuations of network junctions. The theoretical treatments of networks with constrained chains can be roughly grouped into four types: the constrained junction fluctuation model [54,55] in which each network junction is subjected to a domain of constraints; sliplink models [56,57] in which each network chain threads its way through a number of small rings; tube models [58-60] in which each chain is confined within a tube; and primitive path models [61,62] in which the chain segments lie along the shortest path between chain ends.

According to the models, the free energy of deformation and the stress in the network with constrained chains contains an additive contribution to that describing the phantom network. In the junction fluctuation model the free energy of deformation for a perfect network is [54,55)]

$$F = F_{ph} + F_c = \frac{\nu kT}{2} \sum_i (\lambda_i^2 - 1)$$

$$+ \frac{\nu kT}{2} \sum_i \left\{ \varkappa_c \frac{(\lambda_i^2 - 1)(\lambda_i^2/\varkappa_c + 1)}{(\lambda_i^2 + \varkappa_c)^2} - \ln\left[\varkappa_c^2 \frac{(\lambda_i^2 - 1)}{(\lambda_i^2 + \varkappa_c)^2} + 1\right] \right.$$

$$\left. - \ln\left[\varkappa_c^2 \frac{(\lambda_i^2 - 1)(\lambda_i^2/\varkappa_c + 1)}{(\lambda_i^2 + \varkappa_c)^2} + 1\right] \right\} \tag{87}$$

and the tension in uniaxial strain

$$f = f_{ph} + f_c = f_{ph}(1 + f_c/f_{ph}) \tag{88}$$

$$f_c/f_{ph} = [\lambda K_c(\lambda^2) - \lambda^{-2} K_c(\lambda^{-1/2})] (\lambda - \lambda^{-2})^{-1}$$

where F_{ph} and f_{ph} correspond to the contribution of the phantom network and F_c and f_c correspond to the constraints. $K_c(\lambda^2)$ and $K_c(\lambda^{-1/2})$ are functions of the parameter $\varkappa_c$ describing the severity of the constraints. If $\varkappa_c = 0$, then F_c and f_c are equal to zero and Eqs. (87) and (88) go overto those for a phantom network. If $\varkappa_c \to \infty$, corresponding to complete suppression of junction fluctuations, Eqs. (88) are transformed into Eqs. (28) and (29) for the affine network and Eq. (87) to Eq. (30).

Flory [55)] has suggested that

$$\varkappa_c = I\langle r^2\rangle_0^{3/2} (\mu_c/V^0) \tag{89}$$

where I is a universal interpenetration parameter, and μ_c is the number of junctions in the volume V^0 of the state of reference. Using Flory's definition of the tension-temperature coefficient [53)] d ln V^0/dT = 3/2d ln $\langle r^2\rangle_0/dT$ one can arrive at

$$d\varkappa_c/dT = \langle r^2\rangle_0^{3/2} (\mu_c/V^0) (dI/dT). \tag{90}$$

We see that, only if the parameter I is temperature independent, the entropic and energetic components of real networks with the sterical restrictions are identical to that of the phantom or affine network.

The free energy of deformation in the sliplink model is [57)]

$$F = \frac{N_c kT}{2} \sum_i \lambda_i^2 + \frac{N_s kT}{2} \sum_i \left[\frac{(1 + \eta_s) \lambda_i^2}{1 + \eta_s \lambda_i^2} + \log(1 + \eta_s \lambda_i^2)\right] \tag{91}$$

where N_c and N_s are the number of ordinary crosslinks and sliplinks, correspondingly, and η_s is a parameter which measures the ability of a sliplink to slide along the chain passing through it; $\eta_s = 0$ corresponds to the phantom network. The free energy of deformation is attributed solely to entropy change. If the parameter η_s is temperature independent, then this version of the sliplink model corresponds to the ideal entropy-elastic model. If η_s depends on temperature, the entropy contribution is

$$\left(\frac{T\,\Delta S}{W}\right)_{V,T} = \frac{f_s}{f} = -\frac{N_c \sum_i \lambda_i^2 + N_s \sum_i \frac{\lambda_i^2}{1+\eta_s \lambda_i^2}\left[\frac{1+\eta_s}{1+\eta_s \lambda_i^2} + 0.434\right]\frac{d\eta_s}{dT}}{N_c \sum_i \lambda_i^2 + N_s \sum_i \left[\frac{(1+\eta_s)\lambda_i^2}{1+\eta_s \lambda_i^2} + \log(1+\eta_s \lambda_i^2)\right]}. \tag{92}$$

We see that in this case the entropic and the energetic component of the free energy is dependent on deformation and that the energy contribution is of intermolecular origin.

The Marrucci random tube model [58] leads to the following expressions for the free energy and the elastic force at simple extension or compression for tubes with a circular cross section

$$F = \nu k T \frac{l_0}{a_0}\left[\frac{\lambda^2 + 2\lambda^{-1}}{3} + r\left(\frac{\lambda^2 + 2\lambda^{-1}}{3}\right)^{1/2}\right] \tag{93}$$

$$\frac{f}{S_0} = \nu k T \frac{l_0}{a_0}\left[\frac{2}{3} + \frac{r}{3}\left(\frac{\lambda^2 + 2\lambda^{-1}}{3}\right)^{-1/2}\right](\lambda - \lambda^{-2}) \tag{94}$$

where l_0 and a_0 are the tube length and the tube radius of the initial uncrosslinked polymer, respectively, r is the ratio of tube cross-sectional area before and after curing (r = 1 for a perfect network and r < 1 otherwise) and S_0 is the initial cross section of the sample. One may suggest that both l_0 and a_0 have an identical temperature dependence in isotropic networks and, therefore, the ratio l_0/a_0 seems to be temperature independent. Since r is also a constant, we conclude that the tube parameters of the Marrucci tube model are temperature independent. The Gaylord tube model [59] includes also two parameters having the same meaning as in the Marrucci model. Thus, although the tube models give rise to expressions for the elastic free energy which are able to predict some nonclassical effects, the thermoelastic behaviour of the models seems to be similar to the classical entropy-elastic model.

Priss [60] proposed another tube model and obtained the following expression for the elastic force at simple elongation

$$f = 2C_1(\lambda - 1/\lambda^2) + 2C_c[-1/\lambda^2 + (3/2)\lambda/(\lambda^3 - 1) \\ + (\lambda^4 - 4)/2(\lambda^3 - 1)^{3/2}]. \tag{95}$$

where

$$C_1 = (\nu k T/2)\langle r^2 \rangle/\langle r^2 \rangle_0; \quad C_c = \frac{\nu k T K_0^2}{32\alpha_m^2}\left[\frac{(m+1)\,\text{sh}\,2\delta - \text{ch}\,2\delta - 3}{\text{sh}^4\,\delta}\right]$$

$$\alpha_m = 3(m+1)/2Nl^2; \quad \delta = \text{arcch}\,(1 + K_0/2\alpha_m).$$

N is the number of statistical segments in a chain, l is their length and m the number of submolecules. This model includes the front factor only in C_1, but the parameter K_0 is also temperature dependent; it depends on the thermal expansion. Thus, the Priss tube model predicts different temperature dependences of C_1 and C_c. It means that the entropy and energy contribution have to be dependent on λ and include a considerable intermolecular part.

The free energy and the elastic force for simple elongation or compression in the primitive path model is [62)]

$$F = \frac{\mu_c kT}{2}\left[J^2 \frac{(1-\alpha_p)}{(1-\alpha_p J)} + \log\frac{(1-\alpha_p J)}{(1-\alpha_p)}\right] + \mu_c kT[J^2 + \beta_p(J-1)] \quad (96)$$

$$f = \mu_c kT\left[\frac{2}{3} + \frac{\beta_p}{3J} + \frac{(1-\alpha_p)}{3(1-\alpha_p J)} + \frac{\alpha_p J(1-\alpha_p)}{6(1-\alpha_p J)^2} - \frac{\alpha_p}{6J(1-\alpha_p)}\right](\lambda - \lambda^{-2}) \quad (97)$$

where $J = \left(\frac{\lambda^2 + 2\lambda^{-1}}{3}\right)^{1/2}$, μ_c is the number of crosslinks, α_p and β_p are parameters independent of deformation but dependent on temperature and density. Only isoenergetic chains are considered. If α_p and β_p are equal to zero, Eqs. (96) and (97) transform to Eqs. (28) and (30) for isoenergetic chains. Using the definition of the entropy and energy components of the free energy, one can demonstrate that both the components are dependent on λ and that the energy contribution again is of intermolecular origin.

Thus, this consideration shows that the thermoelasticity of the majority of the new models is considerably more complex than that of the phantom networks. However, the new models contain temperature-dependent parameters which are difficult to relate to molecular characteristics of a real rubber-elastic body. It is necessary to note that recent analysis by Gottlieb and Gaylord [63)] has demonstrated that only the Gaylord tube model and the Flory constrained junction fluctuation model agree well with the experimental data on the uniaxial stress-strain response. On the other hand, their analysis has shown that all of the existing molecular theories cannot satisfactorily describe swelling behaviour with a physically reasonable set of parameters. The thermoelastic behaviour of the new models has not yet been analysed.

3 Experimental Methods

Three types of measurements are usually used for the study of thermomechanical behaviour of polymers: the temperature changes resulting from the instantaneous loading of the sample, the temperature dependence of the stress or force and direct calorimetric measurements of heat effects in various deformation modes. Detailed discussion of the first and second types of thermomechanical measurements may readily be found in the literature and, therefore, we confine ourselves to a brief description of typical procedures. As to the third method, we will describe it in

Thermomechanics of Polymers

more detail since it is the main source of information on the thermomechanical behaviour of polymers.

3.1 Measurements of the Temperature Changes (Isoentropic Measurements)

The temperature changes on deformation of condensed matters can simply be measured by thermocouples. The method was first employed by Joule in the 19th century to measure the temperature changes accompanying the rapid, essentially adiabatic, deformations of rubbers and metallic wires and since then it has been widely used for the same purpose. Using conventional thermocouples, temperature changes of magnitude 0.01 K and even smaller can be measured. Two types of thermocouple attachment are used: on the surface of the sample and in the bulk of the sample. Both methods have some shortcomings. The attachment on the surface needs a very good thermal contact between the thermocouple and the sample. The attachment into the body of the sample can change the stress field in the vicinity of the thermocouple and, consequently, the local temperature can change. If the heat capacity of the sample is known, the measured temperature changes may be used to calculate the heat effect resulting from deformation.

Techniques and procedures of such thermoelastic measurements under unidirectional or uniform (hydrostatic) deformation of solid and rubberlike polymers are described in [1, 64–66]. Similar methods have been used more often for recording the temperature changes resulting from the plastic deformation of solid polymers. Besides thermocouples, fluorescent substances, liquid crystals and IR-bolometers are used for such measurements.

3.2 Temperature Dependence of Stresses (Isometric Measurements)

The adiabatic thermomechanical measurement of natural rubber (NR) carried out by Joule showed that the temperature decreased for small strains, but increased for strains above 13–14% extension, which was referred to as the adiabatic inversion point for NR. Although these experimental results agreed rather well with the theoretical predictions, they were disregarded in quantitative studies. They rather played a role of visual evidence of the entropic nature of rubber elasticity. For precise determinations of the entropic and energetic components of the elastic restoring force in rubbers, another approach was found. According to this approach, the total force f is resolved into entropic and energetic components defined by [33]

$$f = f_u + f_s = (\partial U/\partial L)_{V,T} + T(\partial f/\partial T)_{V,L} \tag{98}$$

Experimental determination of the components of the elastic force thus requires measurements of the changes in force with temperature at constant volume and length. The constant volume requires the application of hydrostatic pressure during measurement of the force-temperature coefficient. This experiment is extremely difficult to perform [22, 23].

Instead of measuring the force-temperature dependence at constant volume and length, one can measure this dependence at constant pressure and length but in this case it is necessary to introduce the corresponding corrections. The corrections include such thermomechanical coefficients as isobaric volumetric expansion coefficient, the thermal pressure coefficient or the pressure coefficient of elastic force at constant length [22, 23, 42].

The measurements of the temperature dependence of the restoring force are usually carried out in the temperature range 350 ± 100 K. Determination of f_u requires a rather far extrapolation of experimental results and it restricts the accuracy of this method. This type of thermoelastic measurements requires equilibrium conditions. Most widely, this method is used for simple elongation and seldom for compression [67] and torsion [68, 69].

Although traditionally the thermodynamic treatment of the deformation of elastomers has been centered on the force, the alternative condition of keeping the force (or tension) constant and recording the sample length as a function of temperature at constant pressure is even simpler [23, 27].

This type of thermoelastic measurements is based on the relation

$$\beta_L = -(\partial \ln L/\partial \ln f)_{T,P} (\partial \ln f/\partial T)_{L,P}. \qquad (99)$$

Since in the elastic region $(\partial \ln L/\partial \ln f)_{T,P}$ is always positive, the force-temperature coefficient at constant length and pressure must be of opposite sign to β_L. For rubbers, at some extensions the isometric inversion $[(\partial \ln f/\partial T)_{L,P} = 0]$ must occur since β_L of the isotropic sample is always positive. For solids, such measurements correspond to the determination of the coefficient β of an elastically stretched sample which, however, does not differ from the usual coefficient of thermal expansion.

3.3 Deformation Calorimetry (Isothermal Measurements)

The method of deformation calorimetry was first developed and applied to polymers by Müller in 1957 [1], who also reported the results of many thermomechanical studies on rubberlike and solid polymers and drew attention to such a thermomechanical approach. In his early investigations, Müller used a gas calorimeter based on the principle of a gas thermometer which measures the change in pressure of a gas in which the sample is deformed. The pressure changes in the gas were caused by temperature changes in the sample during deformation. Heat transport also occurs between the sample and the chamber wall which is maintained at constant temperature. This instrument can be used for the simultaneous recording of the stress-strain dependence at simple elongation and thermal effects accompanying the elongation or contraction of the sample. A rather complex construction of the instrument, its not high enough sensitivity and accuracy did not lead to a wide use of the instrument.

Another type of deformation calorimetry which is based on differential thermometry and uses a flowing gas stream as a heat-transfer medium has been developed by Duvdevani et al. [70]. A test sample fastened in the test chamber is surrounded by the flowing gas stream introduced at a constant controlled temperature and flow rate. An electric heater, placed closely to the sample, causes a slight temperature increase in the flowing gas. Heat evolved or absorbed by the sample during deformation tends to change this predetermined temperature difference. In order to keep this difference constant, the electrical power input to the heater is adjusted, the change in power being proportional to the thermal effect of the sample. A stretching device permits to adopt various stretching modes including deformation at a constant rate and sinoidal deformation. A sensitivity of the instrument is strongly dependent on the gas flow rate and its type. A higher flow rate and turbulence improve the time lag but also increase the noise level. Consequently, the flow rate must be optimized according to specific requirements. At typical measuring conditions, this type of flow-gas deformation calorimeter is capable to record heat rates smaller than 20 mJ/s with a sensitivity limited by the noise level of approximately 0.6 mJ/s. This type of deformation calorimeter is also rather difficult to operate and it has not been widely applied in thermomechanical studies.

At the end of the sixties, Godovsky [64,71] developed a fully automatic deformation microcalorimeter based on the Tiang-Calvet principle for simultaneous recording of thermomechanical behaviour of rubbers and solids (films, fibres) at uniaxial deformation. The device consists of two parts; a microcalorimeter and a mechanical loading system with dynamometric assembly. The differential microcalorimeter includes the working and the reference cells. The temperature difference between the working and reference cells resulting from heat flux from or into the test sample is measured by two thermobatteries, which contain about 800 copper-constantan thermojunction each. The maximum sensitivity of this microcalorimeter is about 2×10^{-7} J/s. The time constant of the empty calorimetric cell is 30 s. The calorimeter is capable of working in the ballistic mode and permits to detect thermal effects of less than 1 s duration. The minimum heat pulse that can be detected in the ballistic mode of deformation is about 4×10^{-4} J. The use of the ballistic regime in measurements allows one to analyse two consecutive processes, one of which is of a ballistic type, for example, rapid extension of an elastic body followed by stress relaxation. The dynamometer assembly which includes an automatic bridge-type dynamometer with a sensitivity of 0.5 g/mm allows one to record force and deformation which can be used for determining the work of deformation. Summing up the heat and work, one can obtain the change of internal energy resulting from deformation. This method was widely used by the author of this review.

The above-mentioned method of deformation calorimetry has found a rather wide application. Modifications of the original design were constructed [72-75] and applied for investigating the thermomechanical behaviour of polymers and polymer composites. At the same time, the commercial Calvet-type calorimeters has been used in thermomechanical experiments on rubbers not only in the uniaxial mode [76-78] but also in torsion [79-80]. Thus, deformation calorimetry has proved to be quite adequate in terms of sensitivity, specificity, rapidity and reliability and therefore seems to be the most promising experimental method of thermomechanical type.

A fundamental point to be kept in mind is that most materials and polymers are inelastic. Energy is dissipated during deformation. However, as has been pointed out in the theoretical study of thermodynamics of deformation of solids "inelastic behaviour on the macroscopic scale often may be traced to internal structural changes resulting from deformation. The atomic and molecular processes which govern these changes always require a certain level of activation energy and, therefore, proceed at finite rates which are smaller compared to the rate which statistical equilibrium of active degrees of freedom is established ... There is no distinction between an elastic solid and an inelastic solid in frozen equilibrium" [81]. Hence, it is still of value to consider calorimetric results for mechanically reversible deformations following the basic relations for ideal materials introduced in Sect. 2, since such ideal materials may be quite close to the real ones. This is the main line of our review.

4 Thermomechanics of Molecular Networks and Rubberlike Materials

4.1 Intrachain Energy Effects

According to the theory of rubber elasticity, the elastic response of molecular networks is characterized by two mechanisms. The first one is connected with the deformation of the network, and the free energy change is determined by the conformational changes of the elastically active network chains. In the early theories, the free energy change on deformation of polymeric networks has been completely identified with the change of conformational entropy of chains. The molecular structure of the chains is fully ignored and the energy effects upon deformation of the network may arise only as a result of the compressibility (expansivity) characteristic for usual liquids and is connected with the volume change, i.e. due to changes of the intermolecular interactions upon deformation.

According to the current state of the theory, the deformation of polymeric networks must be accompanied not only by the intrachain conformational entropy changes but intrachain energy changes which depend on the conformational energies of macromolecules. Therefore, reliable experimental determination of these intrachain energy changes and their interpretation by means of isomeric state theory is of fundamental importance for polymer physics.

Numerous investigations concerning the intrachain effects in polymer networks and determination of the contribution of the internal energy have been published in the past 20 years. The main results of these studies have been summarized by Mark [20,21]. The overwhelming majority of the studies have been obtained by measuring the elastic force as a function of temperature under constant pressure and temperature. The interchain effects were thus determined indirectly using corrections of the volume change upon deformation. As has been mentioned above, such a method of determining intrachain energy effects is always connected with a rather far extrapolation of the temperature dependence of the elastic force and requires equilibrium conditions. In recent years, deformation calorimetry has been used widely for determining the intrachain energy contribution to the elasticity of polymeric networks. This method permits to estimate the energy contribution $(\Delta U/W)_{V,T}$ by carrying out measurements at only one temperature at quasiequilibrium conditions. Hence, the question of primary importance is to make comparison of intrachain effects as revealed by classical thermoelastic and thermomechanical (calorimetric) measurements.

The results for two most widely and thoroughly studied networks, namely NR and PDMS listed in Table 1, demonstrate that reliable values of the intrachain energetic effects can be obtained both by thermoelastic and thermomechanical measurements at various deformation modes. These values are also independent of the condition of experiments, i.e. whether the experiments were carried out at constant volume and temperature or constant pressure and temperature. These results permit to make one more very important conclusion: Since they were obtained during independent experiments by various methods on different samples they evidently show a full insensitivity of the energy contribution to the factors which can have an effect on the intermolecular interaction. It agrees with the conclusion made by Mark [84] and Shen and Croucher [29] who have analysed the influence of such factors as degree of crosslinking, crosslinking conditions, type of deformation, extent of deformation and swelling of the networks and have concluded that in the majority of studies all these factors have no influence on the intrachain energy effects. Thus, the conclusions are in full accord with the basic postulate of the Gaussian theory of rubber elasticity — the free energy additivity principle.

One of the main consequence of Eq. (28) must be the independence of intrachain energy contribution on the deformation ratio. The majority of thermoelastic investigations that have been carefully analysed [7,20,29] confirm this theoretical conclusion. However, in some studies dealing with the thermoelasticity of NR, EPR and some other rubbers, a dependence of f_u/f is found at small deformation ratios

Table 1. Energy contribution for NR and PDMS as obtained by various experimental methods

Method	Type of deformation	Experimental conditions (invariants)	$(\Delta U/W)_{V,T} = $ $= f_U/f = $ $= M_u/M$	Ref.
NR				
f — T	Extension	V, L	0.12 ± 0.02	22, 23)
f — T	Extension	V, L	0.23	22, 23)
f — T	Extension	P, L	0.18 ± 0.03	22, 23)
f — T	Extension	P	0.17[a]	20)
f — T	Compression— Extension	P, L	0.18 ± 0.02	67)
M — T	Torsion	P, L, θ	0.17 ± 0.02	69)
Calorimetry	Torsion	P, T	0.20 ± 0.02	79)
Calorimetry	Extension	P, T	0.18 ± 0.01	76, 79)
Calorimetry	Extension	P, T	0.35[b]	82)
Calorimetry	Extension	P, T	0.28 ± 0.03	24, 85)
Calorimetry	Extension (heat inversion)	P, T	0.22 ± 0.03	24)
PDMS				
f — T	Extension	V, T	0.25 ± 0.01	23)
f — T	Extension	P, L	0.27 ± 0.02	83)
f — T	Extension	—	0.2[a]	20)
Calorimetry	Extension	P, T	0.30 ± 0.05	24)
Calorimetry	Extension (heat inversion)	P, T	0.25 ± 0.03	24)

[a] Mean value from the table compiled by Mark [20];
[b] At 45 °C

(see, e.g. Refs. [29, 42]) which is difficult to interpret since it occurs at $\lambda < 1.5$ at which Eq. (28) suitably predicts not only the intrachain energy effects, bur also stress-strain behaviour. Various attempts have been made to explain this dependence, but special studies carried out by Shen [29, 39] have shown that this dependence seems to be a result of the sensitivity of measurements in the low-strain region and experimental error because of the correction term $(\lambda^3 - 1)^{-1}$. This term becomes very large in the region of small deformation. However, this correction does not apply to measurements at constant volume and then the results demonstrate the independence of f_u/f on the deformation ratio at small strains [22, 23].

Until recently, the problem of the dependence of intrachain effects on deformation in the region of low strain was studied exclusively by using thermoelastic measurements. Recently, Godovsky [24, 85] used deformation calorimetry. Typical results are shown in Fig. 3. For all networks studied in the region of small strains $(Q/W)_{P,T}$ and $(\Delta U/W)_{P,T}$ increase sharply but intrachain changes of $(Q/W)_{V,T}$ and $(\Delta U/W)_{V,T}$ are virtually independent of strain including NR and EPR, for which, as mentioned above, a dependence of f_u/f on strain was observed in this region. A large scattering of the values of $(\Delta U/W)_{V,T}$ is seen in Fig. 3 at small strains. Since the term $(\lambda^2 + \lambda - 2)$ tends to be very small at large λ, it can be neglected and, hence, at large λ it is just $(\Delta U/W)_{V,T}$ which is measured.

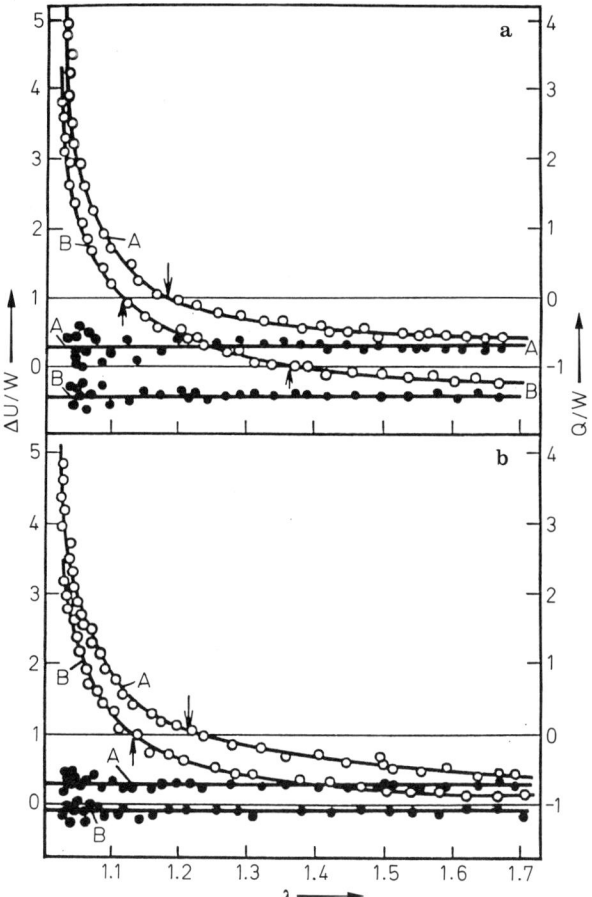

Fig. 3a and b. The calorimetrically determined relative entropy and internal energy contributions as a function of extension ratio λ [24,85]. Room temperature; $\bigcirc$ — $(Q/W)_{P,T}$ and $(\Delta U/W)_{P,T}$; $\bullet$ — $(Q/W)_{V,T}$ and $(\Delta U/W)_{V,T}$. **a** — NR (A), EPR (B); **b** — PDMS (A), PCR (B). The hyperbolic solid lines represent $(Q/W)_{P,T}$ and $(\Delta U/W)_{P,T}$ according to Eqs. (38) and (39) with the following values of parameters α and $d \ln \langle r^2 \rangle_0/dT$ (in K^{-1}): NR 6.6×10^{-4} and 8.2×10^{-4}, EPR 7.5×10^{-4} and -14.3×10^{-4}, PDMS 9.0×10^{-4} and 8.6×10^{-4}, PCR 7.2×10^{-4} and -3.4×10^{-4}. The horizontal solid lines represent $(Q/W)_{V,T}$ and $(\Delta U/W)_{V,T}$ according to Eqs. (33) and (34) with the corresponding values of $d \ln \langle r^2 \rangle_0/dT$

This conclusion permits comparison of the thermomechanical and thermoelastic results for various networks. The most reliable data are summarized in Table 2. The temperature coefficients of the unperturbed dimensions of chains $d \ln \langle r^2 \rangle_0/dT$ as obtained by thermomechanical measurements as well as by viscometry are listed in Table 2. A very good agreement has been obtained. This demonstrates once more the independence of intermolecular interactions of the configuration of the network chains.

Table 2 shows also that for most polymers studied the intrachain energy component may be both positive and negative. According to the isomeric state theory,

Table 2. Comparison of the energy contributions obtained in thermoelastic (f_u/f) and thermomechanical $(\Delta U/W)_{V,T}$ measurements and values of $d\ln\langle r^2\rangle_0/dT$ calculated from the energy contributions and values of viscosity-temperature measurements on isolated chains

Polymer	$(\Delta U/W)_{V,T}$	f_u/f	$d\ln\langle r^2\rangle_0/dT \times 10^3$	
			from	
			f_u/f and $(\Delta U/W)_{V,T}$	Viscosity-temperature
NR	0.21^a	0.18^a	0.66	—
PDMS	0.27^a	0.24^a	0.86^a	0.71 [29], 0.66 [84]
PBR	0.12 [80]	0.11 [20, 84]	0.39	—
EPR	−0.42 [24]	−0.45 [20]	−1.48	—
SBR	~0 [24]	−0.13 (±0.06) [20]	−0.21	—
NBR	0.06 [24]	0.03 [20]	0.15	—
PCR	−0.10 [24]	−0.10 [20]	−0.34	—
PE	—	−0.45 [20]	−1.53	−1.2 [29, 84]

^{a} Mean value from Table 1

positive values of $(\Delta U/W)_{V,T}$ and $d\ln\langle r^2\rangle_0/dT$ are found for polymers for which extended conformations are of higher energy [20, 86]. According to this theory one gets for the conformational energy E_c

$$\left(\frac{\Delta U}{W}\right)_{V,T} = T\frac{d\ln\langle r^2\rangle_0}{dT} = -A\frac{E_c}{kT} \tag{100}$$

where A is a constant depending on chain symmetry and bond angles. Calculations of the trans-gauche energy difference for PE using $(\Delta U/W)_{V,T} = f_u/f = (-0.45)$ led to $E_c = 2$ kJ/mol in excellent agreement with the spectroscopic and thermodynamic data. A summary of the values of energy contribution for many polymers and their interpretation according to the rotational isomeric state theory has been given by Mark [20].

An attempt has been made to correlate the value and sign of the energy contribution with some geometric characteristics of chains, in particular with the cross-sectional area of the chain in the crystalline state [87]. An increase of the cross-sectional area seems to cause a change of the sign of the energy contribution from negative to positive, but physical reasons for such a correlation are unclear.

4.2 Interchain Effects

The statistical theory of rubber elasticity predicts that isothermal simple elongation and compression at constant pressure must be accompanied by interchain effects resulting from the volume change on deformation. The correct experimental determination of these effects is difficult because of very small absolute values of the volume changes. These studies are, however, important for understanding the molecular mechanisms of rubber elasticity and checking the validity of the postulates of statistical theory.

4.2.1 Thermomechanical Inversions

The interchain effects in polymer networks are reflected in the thermomechanical inversion at low strains, which arises from a competition of intra- and interchain changes. Calorimetric studies of unidirectional deformation demonstrates this fact very obviously (Fig. 4). The point of elastic inversion of heat (Table 3) is dependent on the energy contribution and the thermal expansion coefficient in an excellent agreement with the prediction of Eq. (45). The value of $(\Delta U/W)_{V,T}$ for the only one point of deformation, i.e. the inversion point, coincides with data obtained by a more general method (Fig. 3).

An equivalent approach can be found in the determination of the elastic inversion on the f — T dependence [32,34,67], $\lambda_Q = 2\lambda_f$. Data for NR obtained by both methods agree well. The energy contribution in EPR is negative and according to Eq. (37) a thermomechanical inversion of the internal energy should occur which is fully supported by the experimental findings (Fig. 4).

A surprising disappearance of the thermomechanical inversion of heat at elevated temperatures has been observed by Kilian [9,88]. At 90 °C, the thermomechanical inversion in SBR and NR is found to disappear in spite of the constant value of the thermal expansion coefficient. This means that the temperature dependence of elastic force should be negative from the initial deformations, which is in contradiction with experiment. This very unusual phenomenon was supposed to be closely related to "rotational freedom" which will continuously be activated above some characteristic temperature [9,88].

Kilian [9] has also used calorimetric determination of mechanical and thermal energy exchange in isothermal simple elongation for various polymer networks [24] and demonstrated that it can be described by relations which define thermomechanical properties of van der Waals networks (Fig. 4).

The independence of the relative intrachain energy of the extent of deformation permits to resolve the entropy and energy changes into intra- and interchain components. Typical results of this treatment are shown in Fig. 5. It is seen that the prediction of Eq. (43) coincides with the experimentally obtained interchain energy and entropy changes only in the region of low strain ($\lambda < 1.3$). At higher extensions, experimental results are higher and the difference between experiment and theory increases

Table 3. Thermomechanical inversions of heat [24,85,88], internal energy [24,85] and force [89] and related values of energy contribution

Polymer	Heat inversion		Internal energy		Force inversion	
	λ_Q	$(\Delta U/W)_{V,T}$	λ_U	$(\Delta U/W)_{V,T}$	$\lambda_f \approx 1/2\lambda_Q$	f_u/f
NR	1.165	0.22	—	—	1.075	0.17
EPR	1.105	−0.40	1.35	−0.42	—	—
PDMS	1.235	0.25	—	—	—	—
PCR	1.130	−0.08	2.37	−0.10	—	—
SBR	1.150	0ª	—	—	1.070	−0.12ᵇ

ª 30% styrene;
ᵇ 24% styrene

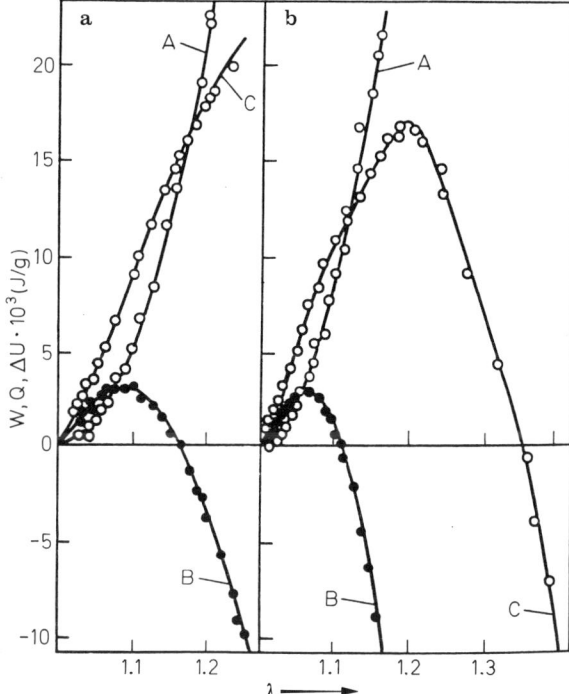

Fig. 4a and b. Mechanical work (A), heat (B) and internal energy change (C) on stretching samples from the unstrained state to λ at room temperature [24, 85]. **a** — NR; **b** — EPR. The solid lines have been computed using Eqs. (35)–(37) with the following values of parameters: NR — C = 0.39 J/g, d ln $\langle r^2 \rangle_0$/dT = 8.2 × 10^{-4} K^{-1}; α = 6.6 × 10^{-4} K^{-1}. EPR — C = 0.61 J/g, d ln $\langle r^2 \rangle_0$/dT = −14.3 × 10^{-4} K^{-1}, α = 7.5 × 10^{-4} K^{-1}. The experimental data can also be represented by Eq. (79) with the corresponding set of parameters [50]

with increasing extension. Such deviations are also found for thermoelastic measurements [23].

An empirical equation which describes the interchain changes $(\Delta U)_{\Delta V}$ and $(\Delta S)_{\Delta V}$ on strain for the four networks has the form [24, 85]

$$(\Delta U)_{\Delta V} = T(\Delta S)_{\Delta V} = C\alpha T \frac{(\lambda - 1)}{\lambda} [1 + \gamma'(\lambda^2 + \lambda - 2)]. \tag{101}$$

Parameter γ depends on the chemical structure of the polymer chain and is equal to 0.1–0.2. The above equation is equivalent to Eq. (57) if γ = 2γ. Equation (101) will be discussed in the next Section.

4.2.2 Strain-Induced Volume Dilation

Interchain changes arising from the deformation of polymer networks are a result of strain-induced volume dilation. Typical results obtained by various methods (Fig. 6) demonstrate that the statistical theory predicts the volume dilation and

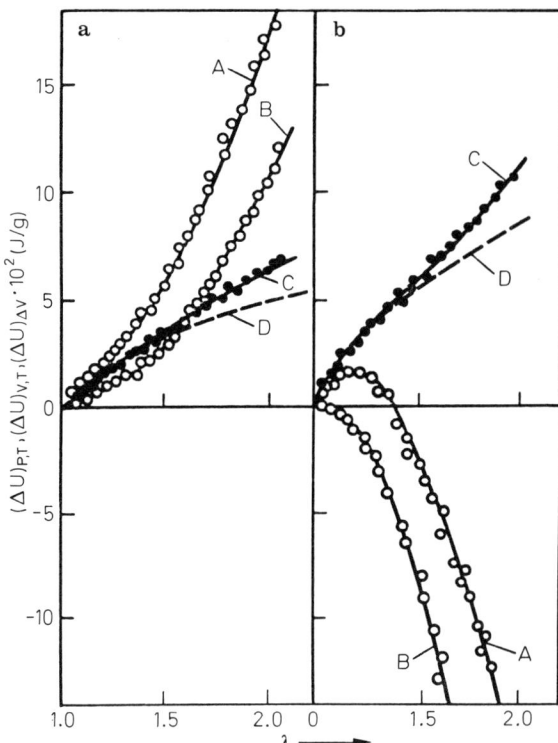

Fig. 5a and b. Intra- and interchain energy changes on stretching samples from the unstrained state to λ at room temperature [24, 85]. A — $(\Delta U)_{P,T}$; B — $(\Delta U)_{V,T}$; C — $(\Delta U)_{\Delta V}$. The solid curves C were calculated using Eq. (101) with $\gamma' = 0.1$ (EPR) and $\gamma' = 0.15$ (NR). D — using Eq. (43)

related interchain energy and entropy changes (Fig. 5) only at λ < 1.3. At larger deformations, experimental results show a significant departure from the statistical theory which underestimates the volume changes on strain.

One of the widely accepted approaches to the analysis of the strain-induced volume changes is the use of Eq. (73) based on the Mooney-Rivlin equation. Although in some studies [23, 29] it has been claimed that the ΔV/V data agree with Eq. (73), Treloar [90] concluded recently that such agreement can not be a rule. He considered the volume change in considerable detail making use of various functional relations for the stored energy W and came to the following conclusions: First, large discrepancies have been found between experiment and the statistical theory. Second, a considerably closer agreement has been obtained for the three-term function of the Ogden type [46], although the discrepancies between experiment and theory exceed 20% at large deformation. It is evident that the apparent agreement between experiment and Eq. (73) obtained on the basis of the Mooney-Rivlin form of W is fortuitous.

In the early treatments of volume changes on elongation of polymer networks it was suggested [25, 34] that the linear compressibility of the strained rubber is isotropic even for moderate deformations. Although this suggestion has been lately refuted theoretically [19], a few experimental data coincide better with this suggestion, which does not mean that the anisotropy of the compressibility is really absent in the strained rubbers. It may only mean that the Gaussian theory does not take into

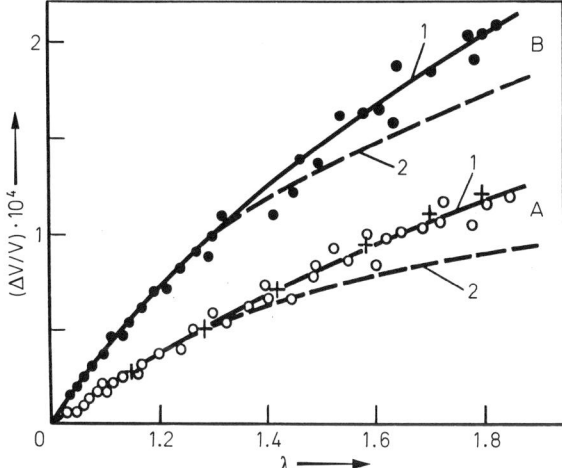

Fig. 6. The relative strain-induced volume dilation of polymer networks as a function of strain. A — NR, B — EPR. ○ and ● — calorimetric data [24, 85]; + — dilatometric data [91]. Solid curves 1 correspond [24] to Eq. (56) with $\gamma = 0.2$ (EPR) and $\gamma = 0.3$ (NR). Curves 1 correspond also to Eq. (84) [50]. Curves 2 correspond to (84) [50]. Curves 2 correspond to Eq. (44) [24]

account some contribution to the volume change. Hence, all these results clearly indicate that neither the statistical theory nor the phenomenological equations of Mooney-Rivlin or Ogden type can predict correctly the volume changes accompanying simple extension of polymer network.

In this connection, it is very interesting that the volume and intrachain changes obtained by various experimental methods [24, 29, 85] [Eq. (101)] agree well with Eq. (56) following from the Tobolsky-Shen semiempirical equation of state or the related phenomenological Eq. (76). The values of γ determined from the data are rather small (0.1–0.3). As has been mentioned above, according to the semiempirical approach by Tobolsky and Shen one can formally suggest that the front-factor in Eq. (28) is pressure dependent. If it is really so, then the parameter γ for rubbers can be considered as an experimental coefficient similar to the coefficient of thermal expansion and compressibility [29].

On the other hand, Kilian [50] having analysed the strain-induced volume dilation [24, 91] using the van der Waals equation of state (Fig. 6) emphasized that only pressure dependence of the interchain parameter, a, is required for a full explanation of the relative volume changes. He arrived at a conclusion that non-crystalline rubbers are anisotropic equilibrium liquids and a higher compressibility of NR was only necessary for fitting the extension data. Hence, on using the van der Waals approach, there is no need of postulating volume dependence of the front factor as proposed by Tobolsky and Shen.

A final comment should be made about the influence of the incorrect prediction of the interchain changes in the Gaussian statistical theory obtained from the results at constant pressure using the corrections following from the theory. The estimations show [24, 85] that the errors of the intrachain effects following from an inadequate description of the interchain changes by the statistical theory are on the level of experimental accuracy of determination of the energy and entropy component. This explains the insensitivity of the components to all experimental factors which can influence the interchain interaction as well as the condition $C_2 \neq 0$ of the Mooney-Rivlin equation.

4.3 Thermomechanics at Large Deformations

In previous sections, we considered the thermomechanical behaviour of Gaussian polymer networks in the region of small and moderate deformations. However, at large deformations approaching maximum extension of chains, non-Gaussian effects arise from the limited extensibility of the network chains and the sharp upturn of the Mooney-Rivlin plot reflects at high deformation the limited chain extensibility [33, 39]. As has been pointed out by Mark et al. [92-95], the observed increases in modulus and the appearance of the upturn of the Mooney-Rivlin curves sometimes may be due to internally generated reinforcement of the network structure arising from strain-induced crystallization or from the formation of hard glassy domains in the case of some block elastomers. The principal effect, resulting from both types of reinforcements, is the generation of "physical" crosslinks which increase the network modulus.

It has been known for a long time that elongation of some networks as NR is accompanied by crystallization [1, 33, 96]. Crystallization is accompanied by a decrease in the internal energy and volume. This decrease is by one or two orders of magnitude higher then that resulting from extension of a network without crystallization. Göritz and Müller [96] have systematically studied strain-induced crystallization in NR by deformation calorimetry as a function of temperature and degree of crosslinking. The degree of extension at which crystallization occurs increases strongly with the temperature. Because of this, one can always choose a temperature at which strain-induced crystallization is absent. During crystallization, $(\Delta U/W)_{P,T}$ becomes negative regardless of the sign of the energy contribution. The limited degree of crystallinity which can be reached is strongly dependent on the extent of deformation. For example, at 15 °C the degree of crystallinity of NR is about 1 % if it is stretched to 300 %, and about 15 % if it is stretched to 400 %. The crystallites formed during elongation are well oriented, with chain backbones parallel to the stretching direction. This conclusion is also supported by investigations of crystallization kinetics of the stretched samples.

In unfilled rubbers, which are not capable of strain-induced crystallization, the upturns on Mooney-Rivlin curves have shown to be absent [92-95]. They disappear also in crystallizable rubbers at elevated temperatures and in the presence of solvents. On the other hand, the upturns do not appear for butadiene, nitrile and polyurethane rubbers if the limited chain extensibility function is introduced in the Mooney-Rivlin expression [97]. Mark [92] has concluded that in the absence of self-reinforcement due to strain-induced crystallization or domains the rupture of the networks occurs long before the limited chain extensibility can be reached.

To support these conclusions, Mark et al. [98-101] have studied elastic properties of model PDMS networks having a bimodal distribution of short and relatively long chains. The short chains are important because of their very limited extensibility. The long chains seem to inhibit the growth of rupture nuclei and thereby make the high elongations possible. Stress-strain isotherms in the temperature range 6–150 °C exhibit the upturns on the Mooney-Rivlin curves which arise from the limited chain extensibility of short chains. The elongations at which the limited chain extensibility become discernible are relatively temperature insensitive, which permits to exclude the strain-induced crystallization. Theoretical calculations have shown that the upturns resulting from non-Gaussian effects begin at approximately 60–70% of the average maximum extensibility of the PDMS chains, which is approximately twice what was generally accepted to be the case [33, 34]. Similar calculations have also shown that network rupture occurs at 80–90% of maximum chain extension.

The general conclusion about crystallization preceding rupture has been lately confirmed by theoretical considerations using non-Gaussian theory of rubber elasticity not only for bimodal networks, but also for short-chain unimodal networks [38]. This new theory has been employed for PE and PDMS short-chain networks with chains having 20, 40 and 250 skeletal bonds. It has been demonstrated that similar upturns in modulus at high deformations must also occur in such networks because of the rapidly diminishing number of configurations consistent with the required large values of end-to-end separations of the chains and, hence, a large decrease in the conformational entropy of the network chains. Calculations have also shown that the increase in modulus of PDMS networks should be significantly different from that of amorphous PE networks having the same number of skeletal bonds and stretched to the same relative length.

Recently, Zhang and Mark [102] carried out the standard thermoelastic (force-temperature) measurements for bimodal PDMS networks from 30 to -50 °C. The constant-length results did not give the expected dependence on λ and $f_u/f = 0.09$ (± 0.02), which is significantly less than the values obtained for unimodal long-chain PDMS networks (Table 2). The difference is supposed to be due to very limited extensibility of short chains. One can suppose that a network chain near its maximum extensibility can no longer increase its end-to-end separation by conformational changes and deformations of bond angles are required. The energies for these deformations are much greater than those for conformational changes and because of that the energy contribution must increase.

Kilian [103] has used the van der Waals approach for treating the thermoelastic results on bimodal networks. He came to a conclusion that thermoelasticity of bimodal networks could satisfactorily be described adopting the thermomechanical autonomy of the rubbery matrix and the rigid short segments. The decrease of f_u/f was supposed to be related to the dependence of the total thermal expansion coefficient on extension of the rigid short segment component. He has also emphasized that calorimetric energy balance measurements are necessary for a direct proof of the proposed hypothesis.

4.4 Thermoelasticity of Liquid Crystalline Networks

In recent years, the behaviour of liquid crystalline polymers including elastomers has been a subject of considerable interest [104, 105]. It is known that small molecule liquid crystals turn into a macroscopic ordered state by external electric or magnetic fields. A similar behaviour seems to occur for liquid-crystalline polymer networks under mechanical stress or strain.

Jerry and Monnerie [106] have proposed a modified theory of rubber elasticity which includes anisotropic intermolecular interactions U_{12} (favoring the alignment of neighbouring chain segments) in the form $U_{12} = \Sigma U_L(r_{12}) P_L(\theta_1) P_L(\theta_2)$, where r_{12} is the intermolecular distance, θ_1 and θ_2 are the angles between the molecular axes and the symmetry axis of the medium, $P_L(\theta)$ are Legendre polynomials. Thus, such interactions are described in the mean field approximation by intermolecular potential having the same form as that usually used for the study of nematic phases. It is shown that the U_{12} interactions increase orientation of chain segments but do not significantly alter the stress-strain behaviour of a network. Although some stress-optical coefficient data in the literature are consistent with the presence of nematic-like interactions, the birefringence measurements can be strongly perturbed by anisotropic internal field effects. The authors believe that linear dichroism and fluorescence polarization could bring more conclusive information.

Rusakov [107, 108] recently proposed a simple model of a nematic network in which the chains between crosslinks are approximated by persistent threads. Orientational intermolecular interactions are taken into account using the mean field approximation and the deformation behaviour of the network is described in terms of the Gaussian statistical theory of rubber elasticity. Making use of the methods of statistical physics, the stress-strain equations of the network with its macroscopic orientation are obtained. The theory predicts a number of effects which should accompany deformation of nematic networks such as the temperature-induced orientational phase transitions. The transition is affected by the intermolecular interaction, the rigidity of macromolecules and the degree of crosslinking of the network. The transition into the liquid crystalline state is accompanied by appearance of internal stresses at constant strain or spontaneous elongation at constant force.

Thermoelastic and photoelastic behaviour of liquid crystalline polysiloxane networks containing the mesogenic side molecules has been studied lately [109]. On cooling, the networks are transformed from rubbery first to nematic liquid crystalline state at T_c and on further cooling into the anisotropic glassy state. At constant force below T_c, the length of the sample increases continously, but it decreases as usual below T_c. At constant length of the sample, the stress decreases rapidly below T_c and approaches zero for small elongations of the sample. For unidirectionally elongated samples, the values of CT (C is the stress-optical coefficient) strongly increase some 20 K above T_c and level off well above T_c. The strong increase of CT is due to a partial orientation of the mesogenic molecules, which is confirmed by X-ray diffraction measurements.

Hence, we are still at the very beginning of thermoelastic and thermomechanical investigations of liquid crystalline elastomers.

4.5 Thermomechanical Behaviour of Rubberlike Materials

Improvement of the mechanical properties of elastomers is usually reached by their reinforcement with fillers. Traditionally, carbon black, silica, metal oxides, some salts and rigid polymers are used. The elastic modulus, tensile strength, and swelling resistance are well increased by such reinforcement. A new approach is based on block copolymerization yielding thermoelastoplastics, i.e. block copolymers with soft (rubbery) and hard (plastic) blocks. The mutual feature of filled rubbers and the thermoelastoplastics is their heterogeneous structure [110].

In spite of their long history, reinforcement mechanisms and elastic properties of elastomers remain the subject of numerous experimental investigations [111-116], but there is only a small number of investigations concerned with measurements of thermoelastic and thermomechanical behaviour. In this section we will critically examine the results of those studies which are dealing with the thermodynamic approach to the reinforcement of rubberlike materials.

4.5.1 Stress Softening: Energetics and Mechanism

An important feature of filled elastomers is the stress softening whereby an elastomer exhibits lower tensile properties at extensions less than those previously applied. As a result of this effect, a hysteresis loop on the stress-strain curve is observed. This effect is irreversible; it is not connected with relaxation processes but the internal structure changes during stress softening. The reinforcement results from the polymer-filler interaction which include both physical and chemical bonds. Thus, deformational properties and strength of filled rubbers are closely connected with the polymer-particle interactions and the ability of these bonds to become reformed under stress.

Numerous investigations [115, 116] offer different viewpoints concerning the mechanism. In one case, this effect is explained by irreversible rupture of the polymer-filler bonds and some overloaded chain segments which lowers the crosslinking density. A large part of the stored energy W is absorbed during the first extension and spent on the cleavage of bonds. According to another viewpoint, stress softening is caused by chain slippage over the particle surface which leads to a redistribution of the chain lengths. The extra work is spent during the first extension on the chain friction over the particle surface. Alternatively, stress softening is explained by non-affine irreversible displacement of crosslinks and entanglements during deformation exhibiting a hysteresis loop. It is important to emphasize that all these viewpoints suppose physical processes, mainly slippage, to be operative on the surface of filler particles.

The stress-strain behaviour of thermoelastoplastics is as a rule a nonlinear one [117, 118]. It strongly depends on many factors, the most important being the volume

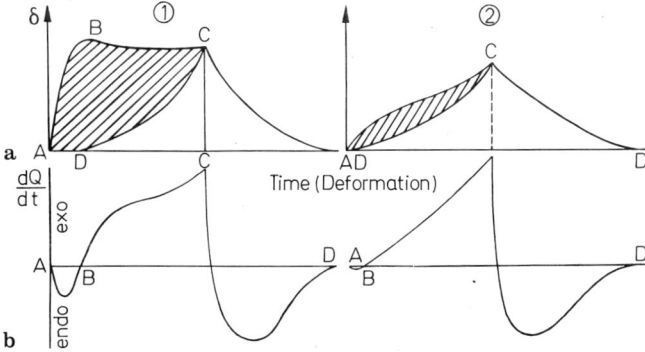

Fig. 7a and b. Scheme of the thermomechanical behaviour of a well phase-separated thermoelastoplastic. ⓐ Stress-strain (or time) curves. ⓑ Plots of heat effects versus time. ① First loading (ABC) and unloading (CD) cycle. ② Second loading (AC) and unloading (CD) cycle. The yielding point occurs at B. AD indicates the residual deformation after the first cycle. AB on the dQ/dT-time curve is the endo-effect resulting from the initial small-strain deformation AB [119]

fraction of the rigid phase. At low content of hard blocks (≤20%), the material behaves as a rubber, while at a high content of hard blocks (30–50%) the stress-strain curve in the first loading often exhibits an initial rapid rise in stress followed by a yielding and cold drawing with neck formation (Fig. 7). The initial elastic part of the load-elongation curve is accompanied by absorption of heat and the neck formation and its propagation along the sample is accompanied by liberation of heat. The plastic flow and the neck propagation continues to high elongations where a rapid increase in stress occurs. Unlike the cold drawing of plastics in the first unloading of block copolymers, there is a substantial strain recovery. Thus, after the first loading-unloading cycle, the initially plastic specimen becomes rubberlike and during the second cycle no yielding and necking occurs. Thermal effects accompanying the second cycle are similar to those resulting from the stretching of filled rubbers. This stress softening phenomenon observed in the first loading cycle of thermoelastoplastics was called the strain-induced plastic-to-rubber transition [120,121].

Quantitative thermomechanical investigations of stress softening were carried out lately on silicon rubbers [7,119,122] and SBS thermoelastoplastics [119,123]. The results obtained reveal some important difference between stress softening of the filled rubbers and thermoelastoplastics (Fig. 8). The mechanical hysteresis for both materials depends on deformation in a similar way. The entire hysteresis loop for filled rubbers is converted into heat. The calorimetric data are in agreement with the current concepts explaining stress softening of filled rubbers by molecular mechanisms such as slippage of the chain over the filler particles [114-116] by which the structure of the rubber is transformed without any change of internal energy.

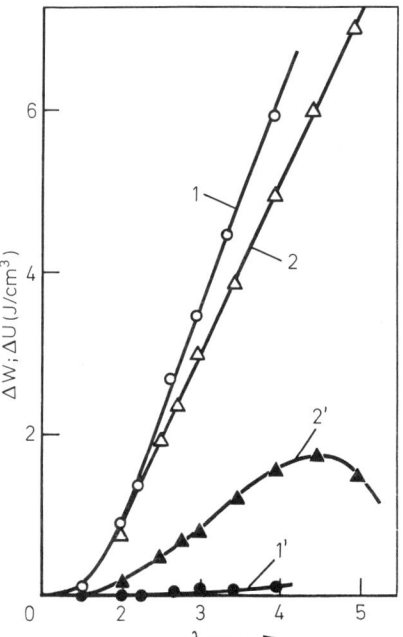

Fig. 8. Dependence of the mechanical hysteresis ΔW (1, 2) of loading-unloading cycle and the change of internal energy ΔU (1', 2') on deformation [119]. 1, 1' — filled PDMS rubber [122]; 2, 2' — star-SBS thermoelastoplastic [123]

On the other hand, stress softening of thermoelastoplastics is accompanied by a considerable internal energy increase. Calorimetric and other results [119] seem to suggest a different mechanism of stress softening in filled rubbers and block copolymers. In filled rubbers, it includes mainly chain slippage and breakage while breakdown and slippage of filler particles has a minor effect. From studies of stress softening of block copolymers it has been concluded that it results from the breakage of a more or less continuous stress-supporting rigid phase and the considerable increase of internal energy is presumably due to this breakdown. An analysis of structural changes and the energy effects has shown [119] that absorption of energy results from plastic deformation of the glassy phase and from appearance of a new surface after breakage of this phase.

The most striking feature of the stress softening phenomenon of thermoelastoplastics is its complete reversibility under certain conditions, consisting in a reformation process in the stress-free state which involves healing of cracks and reaching the initial integrity of the hard phase [119, 120, 124]. The reformation kinetics in stress-softened samples indicates that this process is controlled by diffusion.

4.5.2 Energy Contribution

4.5.2.1 Filled Rubbers

The reinforcement of filled rubbers is usually determined by the particle size and the surface characteristics of filler particles [111-116]. Recent studies have emphasized an important role of internal energy effects in reinforcement [29]. Hence, thermomechanical measurements provide a very important approach to the study of such reinforcement.

In the presence of reinforcing fillers, the elasticity modulus of the elastomers increases in first approximation according to the Guth-Smallwood equation [111-117]

$$E' = E(1 + 2.5\varphi + 14.1\varphi^2) \tag{102}$$

where E is the modulus of elasticity of unfilled elastomer and φ is the volume fraction of the filler. Thus, for filled elastomers, Eqs. (35)–(37) should include parameter φ. Nevertheless, according to Eqs. (33), (34) and (38), (39), the entropy and energy contribution must not be dependent on the presence of the filler. This is a consequence of the intrachain character of these contributions in elastomers, because the filler cannot evidently change the energetics of conformational states of chains. It has been suggested that the role of filler consists in increasing the active chains in network [factor v in Eq. (28)] due to network-filler links. In this connection, the determination of the energy contribution in filled elastomers and its comparison with the values obtained for unfilled networks is of great importance.

The results are shown in Fig. 9. A small amount of the filler strongly increases the energy contribution which is in full contradiction to the assumed increase in the concentration of active network chains caused by the filler. Curve 2 summarizes the results for filled PDMS rubber and for PDMS block and graft copolymers. It is seen that below 20% of the filler or hard phase, the energy contribution is practically independent of the amount of hard phase, but then a considerable increase of $(\Delta U/W)_{V,T}$ is observed. Although in all these cases the energy contribution is

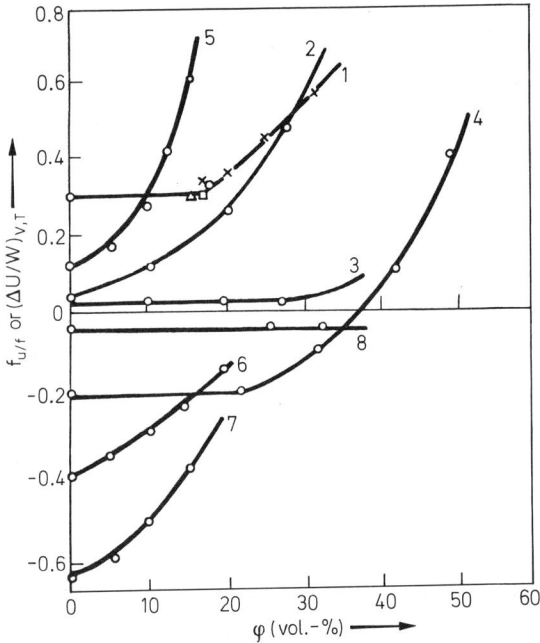

Fig. 9. Dependence of the energy contribution on the filler (filled rubbers) or hard phase (thermoelastoplastics) content. 1 — filled silicone rubber [122] Sil-51 (△), Sil-4600 (□); multiblock copolymer polyarylate-PDMS (○) [128]; graft copolymer of PDMS and AN (×) [128]. 2 — Butyl rubber with high abrasion furnace black [125]. 3 — Butyl rubber with medium thermal black [125]. 4 — SBR-filled rubber [126]. 5 — aerosil *HiSi*-filled silicon rubber [138]. 6 — EPR-filled rubber [129,130]. 7 — plastisized PVC filled with aerosil [131,132]. 8 — SBS block copolymers [134]

dependent on the presence of filler, like in unfilled networks, it is independent of deformation.

The dependence of the energy contribution on the filler amount and its reinforcing ability demonstrates that in filled elastomers the energy contribution seems to loose its obvious meaning as a measure of intrachain effects. Galanty and Sperling [127] supposed that the energy contribution in filled elastomers includes the energy of deformation of hydrogen bonds, movement and deformation of filler particle and association or crystallization of chains of the filler surface. To take this into account, it is necessary to introduce some correction. Galanty and Sperling made such an attempt by introducing into the Tobolsky-Shen equation [Eq. (54)] a factor which reflects a reinforcing ability of the filler and arrived at the following expression for the energy contribution

$$\frac{f_u}{f} = \left(\frac{\Delta U}{W}\right)_{V,T} = T \frac{d \ln \langle r^2 \rangle_0}{dT} + T \frac{d \ln F'}{dT} - \frac{E_f}{RT} \quad (103)$$

where F' is a parameter taking into account the influence of the filler and E_f accounts for the sum of the internal energy contributions by the polymer-filler interaction as well as by the intrachain energy of polymer chain. However, to describe the increase of the elasticity modulus and strength of the filled silicon rubbers, assuming a constant concentration of active chains (parameter v) is impossible.

Therefore, the equations proposed by Galanty and Sperling can hardly be applied to a wide class of filled elastomers. However, regardless of the sign of the intrachain energy changes in a unfilled network, the reinforcement always contributes positively to the energy (Fig. 9) due to intermolecular interactions. Hence, these results demonstrate an important role of intermolecular energy effects in filled elastomers.

4.5.2.2 Block and Graft Copolymers

Stress-softened thermoelastoplastics are another class of reinforced rubberlike materials. The stress softening is a consequence of an extensive fragmentation of the initial stress-supporting hard phase during the first extension. Three problems draw the attention to thermodynamics of deformation of thermoelastoplastics: the energy contribution to the elasticity of a rubbery matrix, the limited chain extensibility at high extensions and the role of hard domains in the hardening at high elongations.

SBS thermoelastoplastics posses the ability to undergo large reversible deformations (up to 2000%), being thus a good model for elucidating the mechanisms of rubber elasticity. It is important to note that polybutadienes used in these thermoelastoplastics are not capable of crystallizing upon extension so, that their thermomechanical behaviour can be studied at large deformations. In segmented polyurethanes and polyesters, the soft phase is built of short-chain polyesters which can exhibit the limited chain extensibility.

Calorimetric investigations of the thermomechanical behaviour of thermoelastoplastics have been carried out in recent years [24, 119, 133, 134]. Taking into account the role of hard particles in the elasticity of filled networks and block and graft copolymers [117, 134, 135], the energy contribution is expected to be indendpent of the hard phase content and deformation. For a series of tri- and polyblock copolymers, $(\Delta U/W)_{V,T}$ is independent of the hard phase content below 40% (Fig. 9, Table 4). It is independent of deformation as well. For the PDMS-AN copolymers, the energy

Table 4. Energy contribution in block and graft copolymers [119, 128, 140]

Copolymer	Block arrangement	Soft block structure	Hard block structure and content	$(\Delta U/W)_{V,T}$
1. Butadiene-styrene				
DST 30	SBS, linear	PB	PS; 0.283	−0.03
Kraton 101	SBS, linear	PB	PS; 0.319	−0.03
Solprene 406	SBS, star	PB	PS; 0.394	−0.13
Solprene 411	SBS, star	PB	PS; 0.340	−0.10
2. Isoprene-styrene				
IST 30	SIS, linear	PI	PS; 0.300	0.25
3. Segmented polyurethanes				
Estane 5707	Multiblock	PTMA	MDI BD; 0.380	0.10
Sanprene E-6	Multiblock	PEA	MDI BD; 0.400	0.20
OBU	Multiblock	PB	MDI BD; 0.175	−0.05
4. Siloxane-arylate	Multiblock	PDMS	Ar; 0.180	0.30
5. Siloxane-acrylonitrile graft copolymers	—	PDMS	PAN; 0.170	0.30
			0.240	0.40
			0.360	0.60

Abbreviations: PTMA poly(tetramethylene adipate); PEA poly(ethylene adipate); MDI diphenylmethane diisocyanate; BD butane diol; PAN polyacrylonitrile; Ar polyarylate; OBU oligobutadiene-urethane contribution increases strongly with the hard phase content (Fig. 9). In this case, the tendency is the same as in the majority of filled rubbers.

Kilian [50] describes work and heat effects at very high extensions of SBS linear and star thermoelastoplastics [24] using the van der Waals equation of state with the following set of parameters in Eqs. (81)–(85); $\lambda_m = 8.5$ (star SBS) and $\lambda_m = 12$ (linear SBS); $a_1 = 0.2$; $a_2 = 0.1$ and $\beta = 2.5 \times 10^{-4}$ K^{-1}. He arrived at the conclusion that non-Gaussian effects at high extensions arise primarily due to the limited chain extensibility.

A very interesting thermomechanical behaviour was found for stress-softened segmented polyurethanes and polyesters with a high hard phase content. Stress softening of such thermoelastoplastics is accompanied by a considerable residual deformation (approximately 50% of the total deformation) arising from plastic deformation and orientation of hard domains in the stretching direction. These oriented block copolymers exert complete elastic recovery on extension and can be used as elastic fibres. Bonart [136-138] has studied the orientational mechanisms and thermodynamics of segmented polyurethanes in considerable detail.

Although the intrachin energy contribution of the soft phase in polyurethanes is negative, the internal energy increases upon extension (Fig. 10). Moreover, the thermomechanical behaviour of the segmented block copolymers and a typical

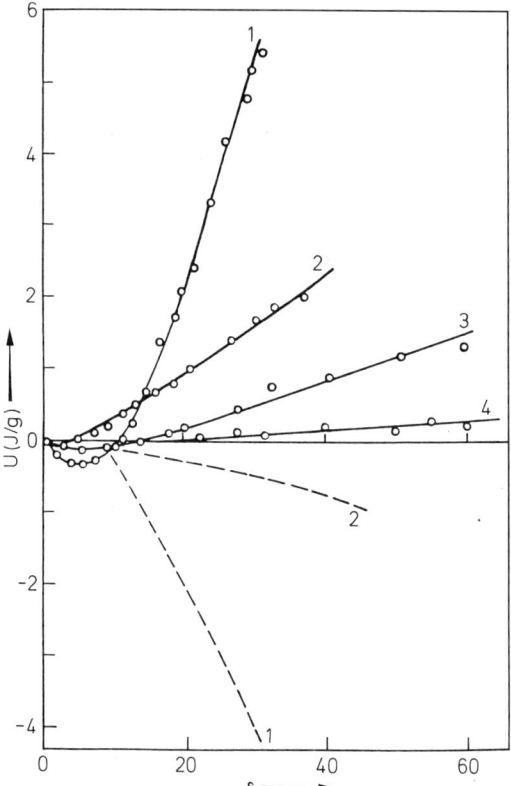

Fig. 10. Internal energy changes as a function of deformation for oriented LDPE (1) and stress softened thermoelastoplastic polyurethanes with 50% (2) and 42% (3) hard phase content and polyether-polyester block copolymer with 48% hard phase content (4). The dotted curves 1 and 2 represent intramolecular energy changes for the corresponding polymers [119]

oriented crystalline polymer is rather similar. These results have strongly supported the idea that the thermomechanical behaviour is controlled not only by conformational changes but also by a change of interchain interaction in the rubbery phase. A further discussion of this idea will be given in Section 5.2.

At high deformation, non-Gaussian effects cause an upturn in the Mooney-Rivlin plot. For the SBS thermoelastoplastics, a similar upturn is observed at high extensions. The upturn for SBS block copolymers can be explained either by the limited chain extensibility or by the presence of glassy incompressible PS domains (cf. Ref. [101]). At moderate deformations, these domains are undeformed, but at high deformations they begin to undergo deformation as is evident by electron microscopy. Hence, the hardening of SBS materials resulting from the deformation of domains seems to be a more real reason for the upturns at high extensions than the limited chain extensibility.

One can expect the limited chain extensibility to be effective in segmented block copolymers because of the lower molecular weights of their soft blocks. However, careful thermomechanical investigations have shown that at large deformations, some additional structuring of hard blocks caused by reformation of hydrogen bonds during deformation occurs [119, 139, 140]. This conclusion is in agreement with the results of Bonart [136-138]. Because of this effect, it is difficult to check whether the intrachain energy contribution is independent of the extension ratio and whether the limited chain extensibility of the soft blocks occurs at large extension. Hence, the problem of approaching the limited chain extensibility in block and graft copolymers remains open and requires further thermomechanical studies.

4.5.2.3 Elastomer Blends

Elastomer blends consisting of two immiscible components are heterogeneous rubber-like materials both components of which are in the rubbery state. Such blends consist usually of either a matrix and a discrete phase or two interpenetrating continuous phases (interpenetrating networks). At homogeneous deformations of such blends, the contribution of either component to the thermomechanical behaviour of the material is determined by the content of the component and the individual characteristics of its chains.

A comprehensive calorimetric study of the thermomechanical behaviour of elastomeric blends was carried out on heterogeneous blends of NR and EPR [141]. These components are completely immiscible and each of them is characterized by a rather large intrachain energy contribution of opposite sign (see Table 2). The energy contribution of the blends at small and moderate deformations, where the strain-induced crystallization of NR is still absent, is an additive function of the composition. The blend containing 60% NR behaves like an ideal rubber since $(\Delta U/W)_{V, T}$ is equal zero. It means that an increase in the conformational energy of NR is totally compensated by a decrease of the conformational energy of EPR. Additivity of the energy contribution has been observed also for some statistical copolymers [29]. In spite of the heterogeneous structure of elastomer blends of all compositions, they behave upon deformation like homogeneous networks.

4.6 Thermomechanics of Bioelastomers

It is well knwon that there exists a number of crosslinked amorphous proteins which in the swollen state exhibit a reversible rubberlike elasticity. Thermomechnical behaviour of elastin has attracted special attention because of its widerspread physiological functions in body tissues [142, 143]. Although elastin has some fibrous structure, it is amorphous and swells in water. Thermodynamics of deformation of elastin was studied using calorimetry and thermoelasticity [143-145]. In a series of microcalorimetric measurements at room temperature in various solvents (water, methanol, ethanol, n-propanol, formamide), it was found that the heat produced during stretching always exceeded mechanical work of stretching. It was almost equal to the heat absorbed during contraction. Because the stress-strain curves were also reversible, the swollen system was treated as a thermodynamically closed one working at constant temperature and pressure. The volume changes resulting from the deformation were not taken into account. From the calorimetric measurements it was concluded that the elastic force arises from interfacial effects.

As has been pointed out in some investigations [146, 147], the calorimetric results have been misinterpreted because no account was taken of the effect of the heat of dilution accompanying the stress-induced increase in swelling. Recent thermoelastic measurements carried out at constant composition (closed system) rather than at swelling equilibrium (open system) over the temperature range 8–35 °C indicated that the elastic force is primarily of entropic origin with the energy contribution $f_u/f = 0.26(\pm 0.09)$ [148]. This value is somewhat larger than that (ca. 0.1) obtained from measurements on open systems [147, 149, 150]. The fact that the elasticity of elastin is primarily entropic in origin and that f_u/f significantly exceeds zero and is of the same order of magnitude as in synthetic elastomers, indicates that the rotational state theory of chain conformations can be successfully used for characterizing thermodynamics of deformation of bioelastomers.

Another area of biology which widely uses the thermomechanical approach is membrane science [151-154].

5 Thermomechanics of Solid (Glassy and Crystalline) Polymers

Experimental studies of the temperature changes and heat effects resulting from reversible deformations of glassy and semicrystalline polymers have been carried out since the 1950s. The early results were summarized and analysed by Müller [1]. Since that time, a number of experimental results on thermomechanical behaviour of glassy and semicrystalline polymers have been obtained. This part of our review is dealing with critical consideration of these results.

5.1 Glassy Polymers

5.1.1 Undrawn Polymers

Three types of measurements were used for investigation of reversible thermomechanical effects in glassy polymers. They include (i) thermoelastic temperature changes at elevated hydrostatic pressure [65, 66], (ii) thermoelastic temperature changes

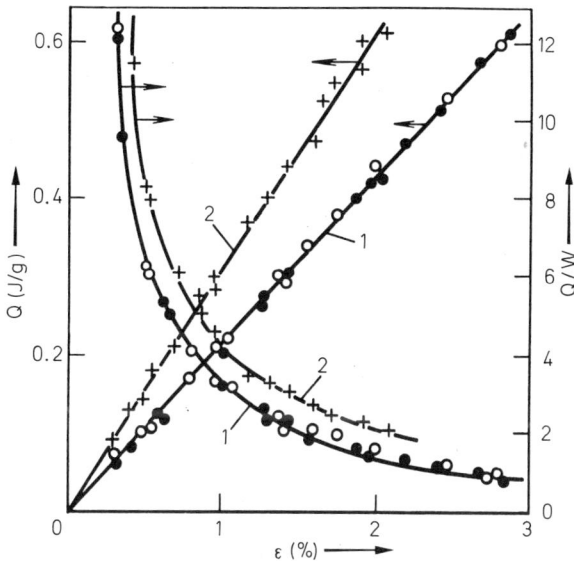

Fig. 11. Heat and heat to work ratio as a function of deformation at 20 °C [7,8]. 1 — PET (○ — amorphousm ● — crystalline w = 50%). 2 — PS. Solid curves were obtained by equations $Q = \beta T E \varepsilon$ and $Q/W = 2\beta T/\varepsilon$ with the following values of parameters: PS — $\beta = 6.8 \times 10^{-5}$ K^{-1}, E = 2.0 GPa; PET — $\beta = 5.4 \times 10^{-5}$ K^{-1}, E = 1.8 GPa

in compression [155] and (iii) calorimetric measurements of heat effects arising from the stretching of films or fibres [7,8,156]. These measurements were carried out on typical glassy polymers such as PMMA, PS, PC, cured epoxy resins and amorphous PET. Quantitative analysis of thermoelastic and thermomechanical effects (ΔT and Q) has been carried out using the Thompson equations [Eqs. (11) and (13)]. The results for two typical glassy polymers (Fig. 11) demonstrate that the polymers behave classically over the reversible stress range, i.e. ΔT and Q resulting from elastic compression or elaongation were a linear function of stress. The linear dependences have been obtained regardless of small stress relaxation (time) effects which occur after a quick straining of the sample by several percent. At small strains that do not exceed 1%, th Deformation was fully reversible for such glassy polymers as PMMA and PS without any hysteresis effects.

The linear thermal expansion coefficient β calculated from these measurements are in excellent agreement with literature data obtained by the conventional method. For example, the values of β calculated from the thermal effects Q during stretching of PS and PET films agree well with conventional dilatometric results, i.e. for PS: $\beta_Q = 6.8 \times 10^{-5}$ K^{-1}, $\beta_{dil} = 7.0 \times 10^{-5}$ K^{-1}; for PET: $\beta_Q = 5.4 \times 10^{-5}$ K^{-1}, $\beta_{dil} = 5.0 \times 10^{-5}$ K^{-1}. The characteristic heat to work ratio η depends hyperbolically on strain which is also in an excellent agreement with prediction following from the thermomechanical analysis (see Fig. 1).

Hence, the results of thermoelastic and thermomechanical measurements demonstrate that typical unoriented polymeric glasses are well described by the classical thermomechanical theory.

5.1.2 Drawn Polymers

The most thorough calorimetric investigation of oriented glassy polymers was carried out by Bonart et al. [156] on PET-films. The effect of degree of cold-drawing of PET-films on the values and sign of thermal effects at small and fully reversible stretching was studied. Typical results are shown in Fig. 12. When the degree of cold-drawing increases, the endoeffect Q resulting from the stretching to the same strain decreases. At degrees of cold-drawing larger than $\lambda = 4$, the heat effect changes its sign and stretching of the film is accompanied by evolution of heat. It means that due to cold-drawing, the linear thermal expansion coefficient changes its sign from positive to negative. In spite of this change, the dependence of heat to work ratio η on strain is hyperbolic, as predicted by the theory (Fig. 1). It is worth mentioning that the samples of a higher degree of cold-drawing than 4 possess some crystallinity, because the cold-drawing occurred near the glass transition. However, as has been shown by Godovsky [7,8], cold-drawing of PET-films at room temperature, which leads only to the appearance of so-called amorphous c-texture without any crystallization, also causes a change in sign. Hence, although stretching of well-oriented PET-films at room temperature (which is well below glass transition) is accompanied by liberation of heat, the strain dependence of η as well as the elastic inversion of the internal energy correspond to the thermomechanical behaviour of the solid with a negative thermal expansion along the stretching axis and is not characteristic for elastomers. The parameters of internal energy inversion and the values of β obtained with the aid of Eqs. (21) and (23) using values of ε_{inv} and ε_{min} are listed

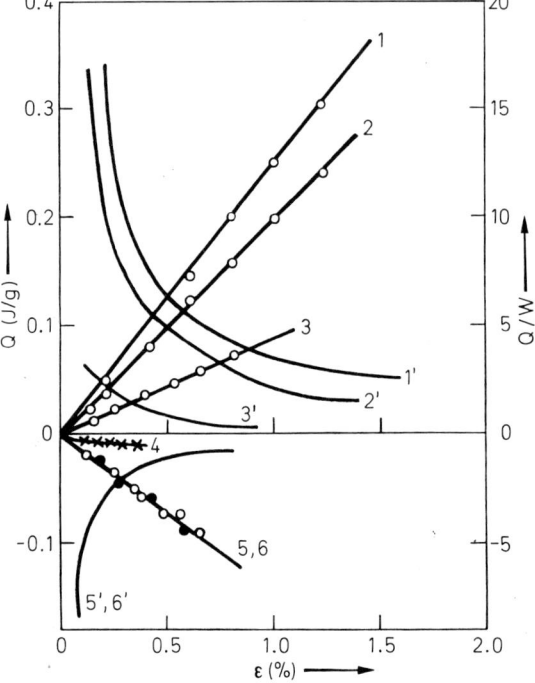

Fig. 12. Dependence of elastic heat (1–5) and heat to work ratio (1′–5′) on strain for PET of various degree of drawing [156]. Degree of drawing: 1, 1′ — undrawn; 2, 2′ — $\lambda = 2$; 3, 3′ — $\lambda = 3$; 4 — $\lambda = 4$; 5, 5′ — $\lambda = 5$; 6, 6′ — $\lambda = 6$

5.1.3 Microphase-Separated Block Copolymers with a Solid Matrix

In Section 4.5.2.2, it has been shown that the first loading of microphase-separated block copolymers with a rather high content of the hard phase often exhibits an initial rapid rise in stress. Thermomechanical investigations [119,157] have shown that such a behaviour is a result of solidlike deformation of the stress-supporting rigid phase. The continuity of the phase seems to depend strongly on the rigid block content and its morphology. Two ways are especially useful for changing the morphology of block copolymers: casting of films from various solvents and processing technique. Figure 13 shows the dependence of β determined from the heat effects in the initial elastic deformation of films and modulus of elasticity E on the solubility parameters δ of the solvent used. These results agree rather well with a general concept of the role of the solvent selectivity in forming a network of rigid glassy domains [110]. It has been well established that the thermodynamic quality of the solvent determines the composition of the coexisting phases and the possibility of formation of a continuous network.

Samples of SBS block copolymers prepared by extrusion or compression molding exhibit a long-range structure consisting of cylindrical PS domains oriented parallel to the extrusion direction, which Keller et al. [158–161] called "single crystal". The specimens are exceptionally anisotropic, i.e. the initial modulus along the cylinder direction is two orders of magnitude higher than that perpendicular to the extrusion direction. Also the thermomechanical behaviour along the two orientation axes differs considerably [119,133].

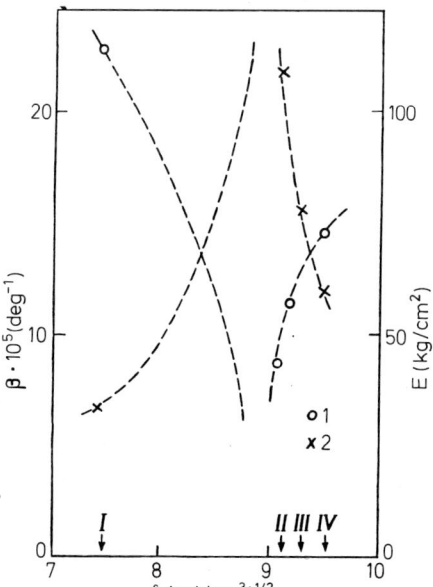

Fig. 13. Effect of the solvent solubility parameter δ on the linear coefficient of thermal expansion β (1) and modulus of elasticity E (2) of films of linear SBS thermoelastoplastics with 28.3% PS obtained from solutions. The solvent is indicated on the abscissa axis: I — n-heptane, II — tetrahydrofurane, III — benzene, IV — chlorobenzene [119]

5.2 Crystalline Polymers

Crystalline polymers are as a rule two-phase systems consisting of both crystalline and amorphous phases and, therefore, their thermomechanical behaviour is more complex than that of glassy polymers, particularly if the amorphous component is above its glass transition temperature. In this case, a two-phase crystalline polymer can be considered as a network in which crystallites are rigid filler particles and act as multifunctional crosslinks. The deformation is localized in the amorphous regions. It has been suggested that, similar to filled elastomeric networks, the elastic properties of such two-phase polymers should be related to the conformational changes in the amorphous regions [162–164]. According to this suggestion, the elastic properties of two-phase crystalline polymers can be analysed using the theory of rubber elasticity. The chains in the amorphous phase are supposed to be in a highly strained state even in the absence of an external macroscopic stress, so that their elasticity has to be considered in terms of the inverse Langevin statistics. The change in the enthalpy on deformation of amorphous regions resulting from the intermolecular changes is supposed to be small and can be neglected. According to this approach, the free energy of deformation of two-phase crystalline polymers above Tg is purely intrachain. However, this approach seems to be incorrect because the changes of interchain interactions in amorphous regions occur during deformation accompanied by a volume change so that the enthalpy changes considerably. The volume changes correspond to that accompanying deformation of solids and not of elastomers. All these facts mean that the free energy of deformation of semicrystalline polymers has an intermolecular origin.

5.2.1 Undrawn Polymers

For characterizing thermal behaviour under elastic deformations of undrawn crystalline polymers, two types of measurements were used; determination of thermoelastic temperature changes during stretching of films and fibres [165–167] and direct calorimetric observation of heat effects accompanying stretching of films and fibres and compression of cylindrical samples [7,8,64]. The measurements were carried out mainly at room temperature on polyolefins, polyamides and polyesters. Absorption of heat occurred on stretching and the return to the initial strain-free state was accompanied by evolution of heat. Such thermal behaviour is qualitatively inconsistent with the suggestion that chains in the amorphous regions are in a highly strained state since the stretching of such chains must be accompanied by evolution of heat. The observed effect is of opposite sign.

As is seen from Fig. 11, the crystallinity does not change the value and sign of thermal effects and heat to work ratio in PET in comparison with a completely amorphous glassy sample. Hence, the thermomechanical behaviour of crystalline polymers at temperatures below T_g is fully identical with that of a completely amorphous glassy polymer.

The thermomechanical behaviour of undrawn semicrystalline polymers above T_g is shown in Fig. 14. The values of the coefficients of thermal expansion calculated from the heat effects agree well with dilatometric results. For PE, the influence of degree of crystallinity on the value of thermal effects and thermal expansion coefficients was also studied [64].

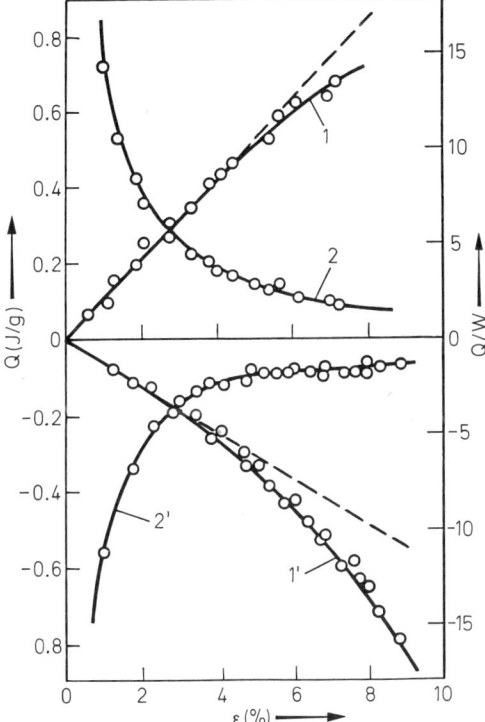

Fig. 14. Heat (1, 1′) and heat to work ratio (2, 2′) for reversible deformation of LDPE at 20 °C [7, 8]. 1, 2 — isotropic sample; 1′, 2′ — cold-drawn sample ($\lambda = 4$, deformation along orientation axis). Dotted curves correspond to the equation $Q = \beta T E \varepsilon$ and solid curve 2 to the equation $Q/W = 2\beta T/\varepsilon$ with $\beta = 2.4 \times 10^{-4}$ K^{-1} and $E = 0.1$ GPa

Most undrawn crystalline polymers possess spherulite morphology with a radial arrangement of fibrils which are complex aggregates of crystallites and amorphous regions.

One can conclude that the presence of crystallites prevents the amorphous chains from deforming exclusively due to rotational rearrangements and the deformation is accompanied by volume changes. For undrawn PE and PP with degrees of crystallinity of 60–70%, the Poisson's ratio is approximately 0.35 at room temperature. This value is typical for solids and it corresponds to the increase of the volume during unidirectional stretching.

From studies of thermomechanical effects in isotropic crystalline polymers, one can finally conclude that the volume elasticity of their amorphous regions is very characteristic of such polymers[1]. This volume elasticity seems to be a consequence of a statistical arrangement of crystallites and amorphous regions and their covalent linkage. Thus, the thermomechanical behaviour of isotropic crystalline polymers contradicts the idea that these polymers can be considered as classical semicrystalline networks and that their elastic properties are determined only by the conformational changes of highly oriented amorphous parts of chains [162–164]. This picture becomes especially evident after considering the thermomechanical behaviour of oriented crystalline polymers.

---

[1] See note at p. 95

5.2.2 Drawn Polymers

Drawn crystalline polymers have turned out to be extremely interesting thermomechanical materials. Already in early studies [1], it was found that cold-drawing of nylon leads to a change in sign of the thermal effect accompanying reversible stretching of a cold-drawn sample, i.e. absorption of heat of an undrawn sample changes to evolution of heat on stretching of a cold-drawn sample. Further comprehensive studies [7,8,64,71,168] have shown that the change in sign is observed in many crystalline polymers. Typical results are shown in Fig. 14 for PE. Quantitative dependences for both undrawn and drawn polymers are similar. The origin of the sign change is determined by the nature of the negative thermal expansivity of drawn crystalline polymers.

5.2.2.1 Negative Thermal Expansivity of Drawn Crystalline Polymers

Above T_g, most drawn crystalline polymers often exhibit a shrinkage on heating, i.e. a negative thermal expansivity along the orientation axis [7,14-17]. Because $\beta_{cr\|}$ is negative (see 2.2.3), one can suggest that this shrinkage is a consequence of that fact. However, a comparison of $\beta_\|$ for drawn crystalline polymers with $\beta_{cr\|}$ indicates that macroscopic $\beta_\|$ may considerably exceed $\beta_{cr\|}$. At least two negative contibutions to $\beta_\|$ must exist. One of them is determined by the negative value of $\beta_{cr\|}$ for crystallites oriented along the orientation axis. The second contribution may be due to the shrinkage of oriented chains in amorphous regions above T_g. Thus, the shrinkage of oriented crystalline polymers on heating seems to include two molecular mechanisms: the conformational elasticity of oriented amorphous chains and the shrinkage of oriented crystallites. In contrast to the isotropic state, in the well-drawn state all crystallites are oriented along the drawing axis and therefore crystallite axes are parallel to the draw direction. On heating they become shorter. This contribution seems to be negative both below and above T_g. The contribution of the amorphous regions is determined by the degree of crystallinity, the portion of the tie chains and the degree of their orientation.

Thus, for understanding the features of the thermal shrinkage of drawn crystalline polymers, it is very important to consider relations typical for stretched elastomers.

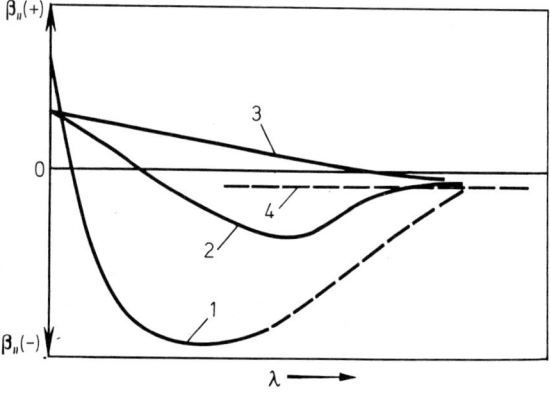

Fig. 15. Hypothetical dependence of the linear thermal expansion coefficient $\beta_\|$ on the degree of orientation [7]. 1 — crystallizable stretched rubber, 2, 3 — crystalline drawn polymers, 4 — $\beta_{cr\|}$

The dependence of $\beta_\|(\lambda)$ on the degree of extension given by Eq. (50) has been observed experimetaly for NR and some other elastomers up to $\lambda \approx 1.5$ [29,169]. At larger extensions, $\beta_\|$ was larger than predicted by Eqs. (50) and (52). In particular, for NR at $\lambda = 1.5$, $\beta_\| = -1.3 \times 0^{-3}$ K^{-1}, at $\lambda = 3$, $\beta_\| = -3.7 \times 10^{-3}$ K^{-1} and at $\lambda = 4.5$, $\beta_\| = -4.2 \times 10^{-3}$ K^{-1}. At $\lambda > 3$, $\beta_\|$ depends only weakly on the degree of extension, and $\beta_\| \to$ const at large λ.

On stretching a crystallizable network, a portion of chains are packed in crystallites their c-axes being oriented along the stretching direction. The thermal shrinkage of macromolecules in crystallites is about 2–2.5 orders of magnitude less than that of chains in the non-crystalline part. Therefore, if the degree of crystallinity increases on extension the thermal shrinkage may decrease (Fig. 15). According to the two-phase model, a stretched semicrystalline elastomer can be represented by a successive arrangement of crystallites and amorphous regions, which leads to

$$\beta_\| = \beta_{am\|}(1 - w) + f_c \beta_{cr\|} w \qquad (104)$$

where f_c is the function of orientation of the crystallites. Because $\beta_{am\|} \gg \beta_{cr\|}$, then at $w < 1$, $\beta_\| \approx \beta_{am\|}$. Thus, a stretched crystallizable elastomer shrinks mainly because of increasing conformational entropy of stretched amorphous chains.

If a stretched elastomer returns to the initial unstretched state at the same temperature, it shrinks to its initial length and the crystallites melt. In contrast, the stresses existing in the amorphous regions of oriented crystalline polymers are balanced by crystallites. Another very important difference from stretched crystallizable elastomers consists in that only some chains in the amorphous regions are the tie-molecules. Let us consider some possible ways of reaching the negative thermal expansivity with degree of drawing [7].

One can suppose that at moderate degrees of cold-drawing of polymers with a low degree of crystallinity, a large portion of chains in the amorphous regions are stretched enough so that $\beta_\| \gg \beta_{cr\|}$ (Fig. 15, Curve 2). On further increasing the degree of drawing, $\beta_\|$ may decrease, which leads to the appearance of a minimum on the dependence on λ. Accepting the idea of rubber-elastic behaviour of amorphous regions of crystalline polymers and neglecting the role of intermolecular interactions, $\beta_\|$ must be always considerably larger than $\beta_{cr\|}$; $\beta_\|$ may approach $\beta_{cr\|}$ only from the region of large negative values. If the conformational mobility is strongly restricted by intermolecular interactions, a second way seems to be possible, namely if $\beta_\| < \beta_{cr\|}$ (Fig. 15, Curve 3) $\beta_\|$ approaches $\beta_{cr\|}$ from the region of small negative values. This is typical for polymers with a large degree of drawing.

Dilatometric studies have demonstrated the negative thermal expansivity for many oriented crystalline polymers [64,170-176]. The results of these experimental studies may be summarized as follows. Cold-drawing of PE below T_m [172] and solid-state extrusion under elevated pressure [170,171] lead to a monotonous decrease of the positive thermal expansion coefficient with increasing draw ratio. At a certain degree of orientation, dependent on temperature, $\beta_\|$ becomes negative with $\beta_\| < \beta_{cr\|}$ (Fig. 16). This is the second way of reaching negative expansivity applied, e.g. to POM (w = 63%, $T_{dr} = 423$ K) [173].

For less crystalline polymers such as LDPE with w = 42%, one gets $\beta_\| = -5 \times 10^{-5}$ K^{-1} at $\lambda = 2$ and $\beta_\| = -20 \times 10^{-5}$ K^{-1} at $\lambda = 4.2$ [173,7] (the first

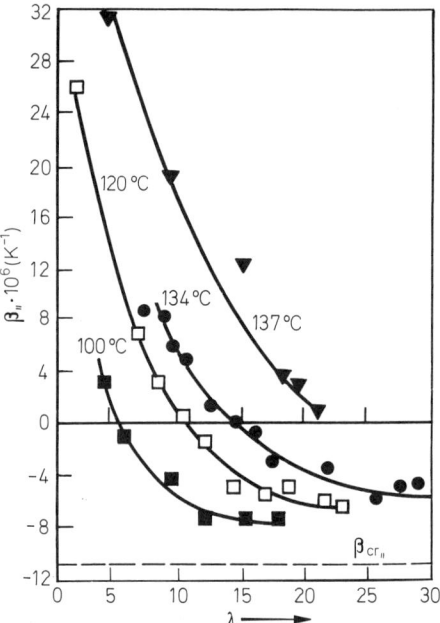

Fig. 16. Dependence of the linear thermal expansion coefficient of HDPE on the degree of drawing at different temperatures [170, 171]

way curve 2, Fig. 15). The same way has been used for polymers with a higher degree of crystallinity, e.g. HDPE with w = 70%, for which at 20 °C $\beta_{\parallel} = -2.4 \times 10^{-5}$ K^{-1} > $\beta_{cr\parallel}$ and at $\lambda = 8\beta_{\parallel}$ is independent of λ [174-176] (cf. also $\beta_{\parallel}$ for HDPE with $\lambda = 10$ and 20 in Table 5). Thus, both ways of reaching the negative expansivity indeed apply for oriented crystalline polymers. In what way the given negative value of $\beta_{\parallel}$ is reached depends on the conditions of orientation, first of all on temperature and degree of drawing. The degree of crystallinity is important because, for polymers with low w, the large drawing cannot be reached.

A very important feature follows from the comparison of $\beta_{\parallel}$ values of oriented crystalline polymers and stretched elastomers and especially of the temperature dependence of $\beta_{\parallel}$. If one estimates $\beta_{am\parallel}$ using Eq. (92) and uses known values of $\beta_{\parallel}$, $\beta_{cr\parallel}$ and w, then $\beta_{am\parallel} = -5.2 \times 10^{-5}$ K^{-1} for the samples studied [174-176]. This value is about two orders of magnitude lower than that for stretched elastomers. The same relation is typical for other high oriented polymers. Such large differences in the negative expansivity of amorphous regions of crystalline polymers and stretched elastomers may be only due to the interchain interaction which prevents oriented chains from a free shrinkage. With increasing temperature, the interchain interaction and internal stresses decrease as a result of their relaxation. This means that $\beta_{\parallel}$ of oriented crystalline polymers must strongly increase with increasing temperature. According to Eq. (50), $\beta_{\parallel}(\lambda)$ is a very weak function of temperature.

The temperature dependence of $\beta_{\parallel}$ (Fig. 17), typical for oriented crystalline polymers, reveals a very large increase of negative values $\beta_{\parallel}$ with temperature. In a rather narrow temperature interval (50–60 K), $\beta_{\parallel}$ increases several times. The temperature coefficient $d\beta_{\parallel}/dT$ does not depend much on polymer type, although the absolute value of $\beta_{\parallel}$ of LDPE is approximately ten times larger than that of other

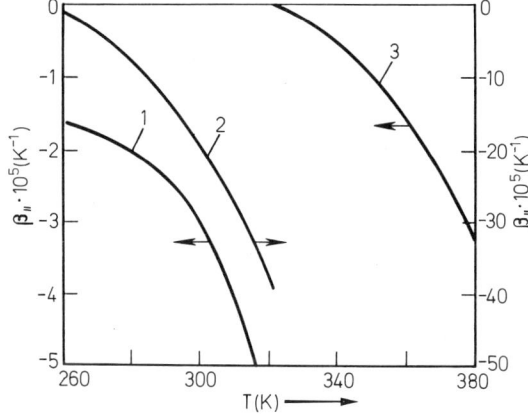

Fig. 17. Temperature dependence of $\beta_{\parallel}$ for drawn crystalline polymers. 1 — HDPE [174-176] (w = 70%, λ = 8–15); 2 — LDPE [14] (w = 43%, λ = 4.2); 3 — POM [14] (w = 63%, λ = 6.5)

polymers. The following fact is also worth mentioning: For many oriented polymers such as POM with λ = 6.5, HDPE with λ = 4.2, PP with λ = 5.5, PA with λ = 2.5, the thermal expansion coefficient $\beta_{\parallel}$ is positive below T_g, increases in the region of glass transition and decreases monotonously on further heating and becomes negative at a temperature characteristic for a polymer and its thermomechanical history [173]. Such a behaviour is consistent with the idea that the intermolecular interaction in amorphous regions is the factor controlling the expansivity and thermomechanical properties of oriented crystalline polymers.

Summarizing the consideration of the thermal expansivity of oriented crystalline polymers, one may conclude that in two-phase polymers the behaviour of their amorphous regions above T_g is controlled by conformational changes strongly restricted by interchain interactions.

5.2.2.2 Thermomechanical Inversion of Internal Energy

The inversion of the internal energy on elastic extension is very important feature of the thermomechanical behaviour of oriented crystalline polymers. Such a behaviour is predicted by Eqs. (20) and (24) for solids having negative thermal expansivity. Such an inversion of the internal energy is observed in oriented crystalline polymers (Fig. 18, Table 5). Although a similar inversion of the internal energy may also occur upon extension in elastomers (4.2.1), the sign of the inversion and its molecular origin is quite different. Indeed, for oriented crystalline polymers, ΔU at first decreases and only then begins to increase. In elastomers, such a change of ΔU cannot be observed in principle. It is important to point out that the inversion of the internal energy in crystalline polymers is determined only by the value of $\beta_{\parallel}$, but the intrachain conformational energy changes do not play any role in this case.

According to Eq. (20) the compression of isotropic solid polymers having positive thermal expansivity must be accompanied by the internal energy inversion. ΔU inversion at compression has been estimated [7] to occur at strains 5–15%. At compression, irreversible plastic deformations occur which prevents a correct experimental determination of ΔU. With inversion parameters, ΔU for isotropic poly-

Table 5. Parameters of internal energy inversion on stretching drawn and compressing undrawn polymers [7, 8]

Polymers	Characteristics of samples[a]	ε_{inv} %	ε_{min} %	ΔU_{min} J/cm³	$\beta_\| \cdot 10^5$ K⁻¹	$E_\|$ GPa	Poisson's ratio
Extension of drawn polymers							
LDPE	$\lambda = 4$; $T_{dr} = 20$ °C	10.5	5.3	−0.43	−19.40	0.27	—
HDPE	$\lambda = 10$; $T_{dr} = 95$ °C	4.5	2.2	−0.40	− 7.50	1.70	0.56–0.63
HDPE	$\lambda = 20$; $T_{dr} = 95$ °C	2.2	1.0	−0.45	− 3.65	8.0	—
PP	$\lambda = 8$; $T_{dr} = 20$ °C	3.4	1.8	−0.39	− 6.30	2.35	0.55
PP	$\lambda = 30$; $T_{dr} = 140$ °C	2.2	1.0	−0.22	− 3.70	3.8	—
PA	Commercial film	5.0	2.4	−0.15	− 8.20	0.55	0.50–0.55
PET	$\lambda = 5.4$; $T_{dr} = 20$ °C $\varrho = 1.338$ g/cm³	0.5	0.2	−0.012	− 0.80	5.0	0.58
PET [156]	$\lambda = 6.0$; $T_{dr} = 85$ °C $\varrho = 1.380$ g/cm³	0.27	0.13	−0.015	− 0.50[b]	14.0	—
Compression of undrawn polymers[c]							
LDPE	$w = 0.40$	−14.4	−7.2	−0.25	24.60	0.1	0.35
HDPE	$w = 0.55$	− 8.5	−4.2	−0.54	14.60	0.6	—
PP	$w = 0.64$	− 6.2	−3.2	−0.32	11.20	0.62	0.35
PS	amorphous	− 4.0	−2.0	−0.38	6.80	2.0	0.33
PET	amorphous	− 3.2	−1.5	−0.22	5.40	1.8	—
PET	$w = 0.40$	− 3.2	−1.5	−0.22	5.40	1.9	—

[a] λ — degree of drawing; T_{dr} — drawing temperature;
[b] Calculated according to Eq. (24);
[c] Values of ε_{inv}, ε_{min} and ΔU_{min} were computed according to expressions given in B.2.2 using β and E obtained in extension and compression (small strain) experiments

mers under compression may be calculated with formulas for inversion at elongation using the values of β and E obtained both for stretching films and compressing small cylindrical samples (Table 5). Finally, it must be pointed out that evolution of heat resulting from reversible stretching is not connected with any crystallization process in amorphous regions and is a thermodynamic consequence of their negative thermal expansivity.

Above T_g, moderately oriented crystalline polymers have an elasticity modulus of the order of magnitude of a few GPa and are capable of being reversibly deformed by 15–25%. Since this deformation is localized in the amorphous regions, the true strain in these regions at typical degrees of crystallinity (30–70%) may be several times higher. The large reversible deformations of amorphous regions with a relatively small elasticity modulus must be due to the transition of unstretched conformations into the stretched ones. Such transitions in oriented crystalline polymers, such as PE, PP and PA, have been recorded by IR spectroscopy [177, 178]. However, in contrast to elastomers, the gauche/trans transitions in amorphous regions of crystalline polymers are accompanied by a simultaneous volume and intermolecular energy change which determine the modulus of elasticity. Therefore, the traditional approach to the elasticity modulus of crystalline polymers, considered as networks reinforced with crystallites, is inadequate for description of the deformation mechanisms. Therefore, the character of the internal energy changes at stretching of oriented crystalline polymers is consistent with the assumption that their thermomechanical behaviour both below and above T_g is controlled by intermolecular interactions.

5.2.2.3 Thermomechanical Behaviour and Morphology of Drawn Crystalline Polymers

Numerous studies of the structure and properties of drawn crystalline polymers have led to the microfibrillar model of fibrous morphology [177,179,180]. According to Peterlin [179] and Prevorsek et al. [180], the long and thin microfibrils are the basic elements of the fibrous structure. The microfibrils consist of alternating folded chain crystallites and amorphous regions. The axial connection between the crystallites is accomplished by intrafibrillar tie-molecules inside each microfibril and by interfibrillar tie-molecules between adjacent microfibrils.

Godovsky [7,8], following Takayanagi's approach of calculating the modulus of elasticity of two-phase polymers, has analysed the thermophysical behaviour of various structural models of drawn crystalline polymers. Here, we will consider only the Peterlin-Prevorsek model which is the most suitable for explaining the thermomechanical behaviour. According to this model, the thermodynamics of stretching is described by Eqs. (17)–(21) with the following expressions for the modulus $E_{||}$ and the thermal expansion coefficient $\beta_{||}$ along the orientation axis

$$E_{||} = E_{AI||}X + \frac{(1 - X)^2 E_{AB||} E_{cr||}}{E_{cr||}(1 - w - X) + wE_{AB}} \quad (105)$$

$$\beta_{||} = \frac{\beta_{AI||} X E_{AI||}}{E_{AI||} X + \frac{(1 - X)^2 E_{AB||}}{(1 - w - X) + w(E_{AB||}/E_{cr||})}}$$

$$+ \left(\frac{(1 - X) E_{AB||}}{E_{AI||} X + (1 - X)^2 E_{AB||}} \right.$$

$$\left. \times \frac{\beta_{AB||}(1 - w) + \beta_{cr||}(w - X)}{(w - X) E_{AB||}/E_{cr||} - (1 - X)} \right). \quad (106)$$

In these expressions, the subscript AB denotes the intrafibrillar amorphous tie-chains, AI the interfibrillar tie-chains and X is the volume fraction of the interfibrillar tie-molecules.

The analysis has shown that $\beta_{AI||}$ may only be negative, and $\beta_{AB||}$ both positive and negative. Therefore, the thermal effect accompanying a reversible stretching of the model depends on the ratio between $\beta_{AB||}$ and $\beta_{AI||}$ and may be a function of strain even at small strains. Besides, Poisson's ratio for such a heterogeneous model may exeed 0.5. Direct measurements of Poisson's ratio for a number of various oriented crystalline polymers are consistent with this suggestion (see Table 5).

The fibrillar structure of crystalline polymers is determined by molecular characteristics, the initial morphology and orientation conditions. Recently, a complex investigation of the effect of molecular parameters (MW, MWD and degree of branching) and orientation parameters (temperature and draw ratio) on the morphology of PE and its thermomechanical behaviour has been reported [181-185].

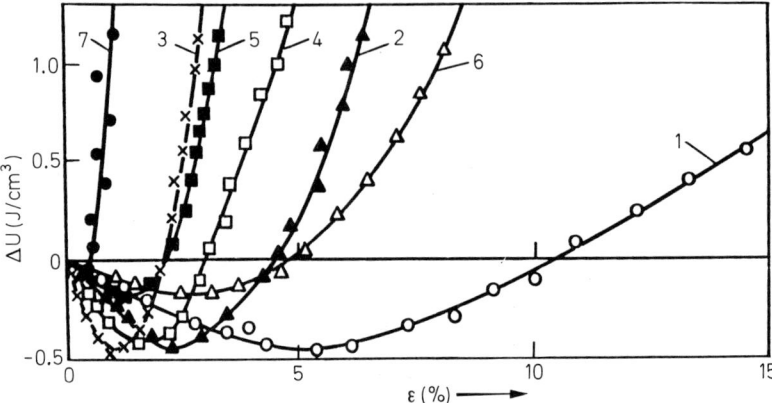

Fig. 18. Internal energy changes ΔU as a function of strain for drawn crystalline polymers at 20 °C [7, 8]. 1 — LDPE ($\lambda = 4$); 2 — HDPE ($\lambda = 10$); 3 — HDPE ($\lambda = 20$); 4 — PP ($\lambda = 8$); 5 — PP ($\lambda = 30$); 6 — PA (commercial film); 7 — PET ($\lambda = 5.4$)

It included a calorimetric investigation of thermomechanics combined with an X-ray spectroscopic analysis of structural changes and the analysis of molecular mobility of PE with the aid of a molecular probe. The sign and values of the thermal effects on stretching depend on the molecular weight and strain (Fig. 19). From the combined analysis of the thermomechanical behaviour and structural changes, it has been concluded that the heat evolution seems to be due to a compression of the interfibrillar amorphous molecules and the heat adsorption to expansion of the intrafibrillar regions conteining a small number of tie-molecules. The resultant thermal effect depends on the relative ratio of the compressed and expanded amorphous regions.

A method was proposed for the quantitative evaluation of the volume fraction occupied in PE by interfibrillar tie-molecules based on the thermal effect and long-period variations in deformation. The method involves summing up the contributions of the thermal effects of the intra- and inter-fibrillar amorphous regions. Estimates have shown that low-molecular-weight specimens were practically free of interfibrillar amorphous regions. An increase in molecular weight causes a monotonous increase of the fraction of interfibrillar regions for comparable temperatures and degrees of drawing. Furthermore, their fraction increases with the degree of drawing at a given temperature and with decreasing draw temperature which is consistent with theoretical estimates [179] and X-ray structural measurements [186]. Thus, the thermomechanical behaviour of oriented PE can be described well in terms of the Peterlin-Prevorsek model.

Annealing of free or fixed oriented crystalline polymers under atmospheric pressure brings about considerable changes in their morphology and, as a consequence, of their mechanical properties [177], [179]. The change are most expressed in ultimately stretched interfibrillar chains and the tie-chains in the intrafibrillar amorphous interlayers. Annealing is usually accompanied by the transformation of the fibrillar morphology into the lamellar one. The strongly oriented specimens

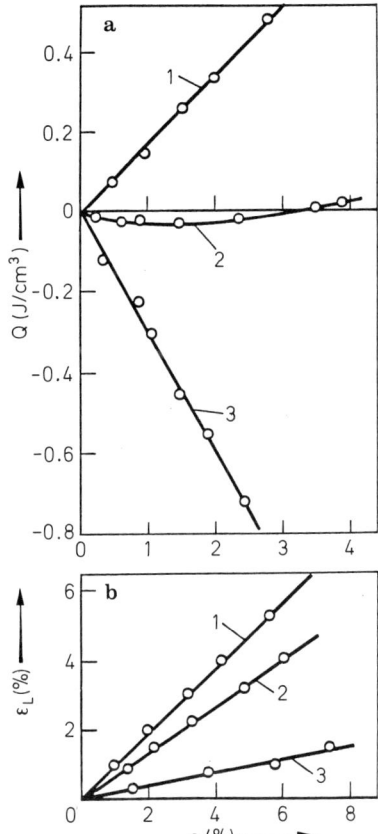

Fig. 19a and b. Heat of elastic elongation (a) and deformation of long period as a function of macroscopic strain (b) for PE [8,181–183]. Sample characteristics: 1 — $\overline{M}_\eta = 4.2 \times 10^4$, $\overline{M}_n/\overline{M}_\eta = 3.0$, $\lambda = 6$. 2 — $\overline{M}_\eta = 3 \times 10^5$, $\overline{M}_n/\overline{M}_\eta = 1.95$, $\lambda = 6$. 3 — $\overline{M}_\eta = 7 \times 10^6$, $\overline{M}_n/\overline{M}_\eta = 2.5$, $\lambda = 5.5$

after annealing in the free state exhibit a change in sign of the extension heat (Fig. 20). The variations may be explained by the concept of two types of amorphous regions. Annealing deorientates and loosens the stretched portion of chains. The expansion of intrafibrillar amorphous regions predominates at low extension and contributes by an endothermal effect. However, the endo-effect becomes reversed as the strain increases. This is particularly apparent in specimens with a lower degree of orientation. Although in isometric annealing the elasticity modulus remains practically unchanged, the thermal effect of deformation and $\beta_\parallel$ vary considerably, and in some cases a change in sign may occur. Therefore, the thermophysical phenomena occurring in deformation of annealed samples reflect closely the changes brought about by annealing.

These observations are valid not only for PE. The changes in the thermomechanical behaviour after annealing of oriented PP having a modulus of elasticity of $E = 10$–15 GPa and PA ($E = 8$–10 GPa) [187] are similar to those of PE. The thermomechanics of the so-called hardelastic fibres, which are known to have a lamellar structure, is very similar to that of annealed PE and PP [188].

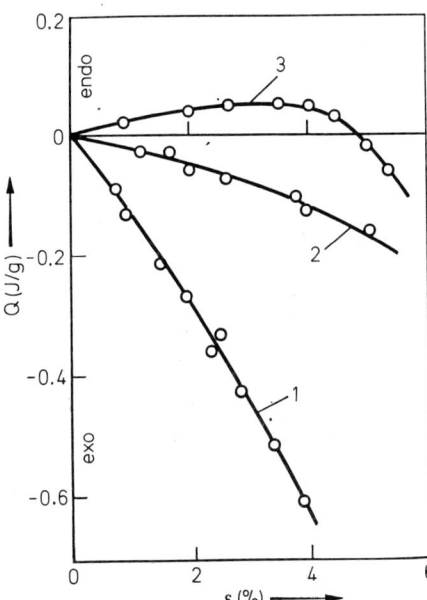

Fig. 20. Effect of annealing of HDPE ($\lambda = 15$) on the heat of elastic extension [8]. 1 — initial unannealed sample; 2 — after isometric annealing at 132 °C for 30 min; 3 — after free annealing at 132 °C for 30 min

The above analysis is not limited to oriented crystalline polymers usually having an elasticity modulus of only a few percent of the theoretical modulus of the crystal lattice. For the highly oriented PP with the elasticity modulus as high as 25–30% of that of the crystal lattice, it is also possible to account for the thermophysical and structural changes occurring in original and annealed samples in terms of the Peterlin-Prevorsek model. Finally, the thermomechanical properties of superoriented super-high modulus PE and PP can also be successfully described in terms of this model [7,8].

5.3 Anisotropy of Thermal Expansivity and Thermomechanical Behaviour

Stretched elastomers and polymer crystallites exhibit a striking anisotropy of the linear thermal expansion and thermomechanical behaviour. A strong anisotropy of these properties is also typical for solid oriented polymers. The linear thermal expansivity decreases along the orientation axis ($\beta_{\parallel}$) and increases in the direction perpendicular to the draw direction ($\beta_{\perp}$) both for crystalline and glassy polymers. However, the anisotropy of oriented crystalline and glassy polymers may be considerably different at the comparable draw ratio. The ratio $\beta_{\perp}/\beta_{\parallel}$ is strongly dependent on the type of polymer. For glassy PS and PMMA at a draw ratio of $\lambda = 4$, the ratio is equal to 1.1 and 2.5, respectively, and for PVC which has a low degree of crystallinity, it is equal to 4.5 at a draw ratio of $\lambda = 3$ [189]. The ratio $\beta_{\perp}/\beta_{\parallel}$ changes considerably with increasing draw ratio [190]. A typical dependence of $\beta_{\parallel}$ and $\beta_{\perp}$ on the draw ratio of crystalline polymers is shown on Fig. 21.

Two different approaches have been used for the theoretical analysis of the anisotropy of thermal expansivity. One is based on the analysis of the effective thermal

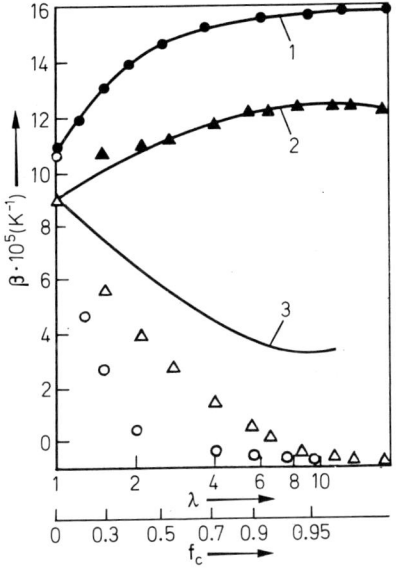

Fig. 21. Anisotropy of linear thermal expansivity of drawn crystalline polymers as a function of draw ratio [172]. ○ — ($\beta_{||}$) and ● — ($\beta_\perp$) HDPE (w = 0.8, T_{dr} = 353 K); △ ($\beta_{||}$) and ▲ ($\beta_\perp$) PP (w = 0.6, T_{dr} = 398 K). Solid curves: 1, 2 — $\beta_\perp$ from Eq. (107), 3 — $\beta_{||}$ from Eq. (108)

expansion coefficients using the methods applied in the theory of thermoelasticity of microheterogeneous materials [191,192]. These methods require the complete compliance tensors of both phases, and of the composite as well, which are usually not available. However, a more essential restriction of this approach is the assumption that in the composite material the properties of the components are constant and identical with their initial properties. Although for two-phase crystalline polymers this suggestion is not valid because of the existence of amorphous regions, some attempts have been made to roughly estimate the expansivity of crystalline polymers [172]. Expressions for the thermal expansivities perpendicular and parallel to the draw axis have the following form

$$\beta_\perp = w\beta_{cr} + (1-w)\beta_{am} + (1/3)wf_c(\lambda)(\beta_{cr\perp} - \beta_{cr||}) \qquad (107)$$

$$\beta_{||} = w\beta_{cr} + (1-w)\beta_{am} - (2/3)wf_c(\lambda)(\beta_{cr\perp} - \beta_{cr||}) \qquad (108)$$

In these equations $\beta_{cr} = (1/3)(2\beta_{cr\perp} + \beta_{cr||})$ is the average coefficient of linear thermal expansion of crystallites and $f_c(\lambda)$ is the distribution function of the crystallites axes. For the isotropic sample $f_c = 0$ and both Eqs. (107) and (108) reduce to

$$\beta = w\beta_{cr} + (1-w)\beta_{am} \qquad (109)$$

From Eqs. (107) and (108), it is evident that the average linear thermal expansion coefficient. i.e. $\beta_{av} = 1/3(2\beta_\perp + \beta_{||})$ or $1/3\alpha$ is independent of draw ratio and equals to the isotropic value β_{iso}. According to this approach, the dependence on λ of the linear thermal expansivites is determined solely by the orientation of the crystallites along the draw direction. When all the crystallites have become completely oriented ($f_c = 1$), a saturation occurs and both $\beta_\perp$ and $\beta_\parallel$ become independent of λ.

This model is incorrect because the linear thermal expansivity for both components in the isotropic and oriented state is assumed to be the same. The concequences of this assumption are quite different for $\beta_\perp$ and $\beta_\parallel$. For $\beta_\perp$, it does not play any essential role, because in the isotropic and oriented state perpendicular to the draw axis the thermal expansivity is determined solely by intermolecular interactions. For $\beta_\parallel$, this suggestion may lead to a principal inconsistency. This conclusion is evident from comparison of the calculated and experimental λ-dependences of $\beta_\perp$ and $\beta_\parallel$ for PE and PP according to Eqs. (107) and (108). For $\beta_\perp$, the agreement between the model calculation and the experiment is quite satisfactory for all draw ratios. On the other hand, Eq. (108) does not describe the λ-dependence of $\beta_\parallel$ at all. This equation does not yield negative values of $\beta_\parallel$ even in case of a limited orientation of crystallites ($f_c = 1$) because it is based on the suggestion that β_{am} is always positive and $|\beta_{am}| > |\beta_{cr}|$.

The other approach concerns the analysis of models of oriented crystalline polymers [7,8,172]. For the Peterlin-Prevorsek model, the expression for $\beta_\perp$ can be represented as

$$\beta_\perp = \frac{\beta_{cr\perp} w E_{cr\perp} + \beta_{AB\perp} E_{AB\perp}(1-w)}{E_{cr\perp} w + (1-w) E_{AB\perp}} (1-X) + \beta_{AL\perp} X . \tag{110}$$

According to Eq. (110) $\beta_\perp$ must only be positive, As has been pointed out above, Eq. (106) allows a negative thermal expansivity. Therefore, the analysis of the structural models of oriented crystalline polymers seems to predict the anisotropy of thermal expansivity more correctly.

Additionally, one can consider the thermal expansivity in an arbitrary orientation direction. For films, the linear coefficient of thermal expansion at angle φ to the orientation axis is determined by

$$\beta_\varphi = \beta_\parallel \cos^2 \varphi + \beta_\perp \cos^2 (90 - \varphi) . \tag{111}$$

Because $\beta_\parallel$ may be negative, there may exist such directions in oriented crystalline polymers along which the thermal expansivity is zero. Thermomechanical studies of oriented LDPE, PA and PET have shown that a reversible extension of the films at an angle of $\sim 30°$ to the orientation axis is not accompanied by thermal effects [64], i.e. $\beta_{30°} \approx 0$. The dependence of β_φ on the angle for LDPE is shown in Fig. 22.

The thermodynamics of deformation of solid-oriented polymers in any direction is determined by expressions which may be obtained from Eq. (22) [7]. The coefficient β_φ may be determined by Eq. (111) and the modulus of elasticity E_φ can be calculated according to the following expression [193]

$$E_\varphi = \frac{E_\parallel}{\cos^4 \varphi + b \sin^2 2\varphi + c \sin^4 \varphi} \tag{112}$$

where $c = E_\parallel/E_\perp$, $b = E_\perp/E_{45°} - (1 + c)/4$.

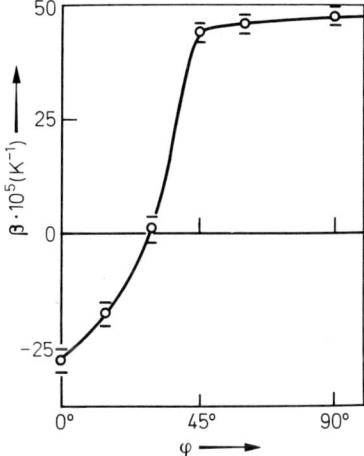

Fig. 22. Dependence of the linear coefficient of thermal expansion of LDPE on the angle to the drawing direction [7]. Values of β were obtained from the heat of elastic stretching.

Experimental studies of the anisotropy of thermomechanical behaviour on LDPE, PA and PET support the idea that in any direction the thermomechanics is controlled by intermolecular interactions [7,64].

5.4 Filled Solid Polymers

Fine solid fillers markedly affect not only the physical and mechanical properties of elastomers but of solid polymers as well. A study of the interaction of a glassy plasticized PVC-matrix with reinforcing (silica) and inactive (chalk) fillers by deformation calorimetry was made recently [131]. Filling of PVC with silica is accompanied by a considerable increase of W, Q and U on stretching. The same amount of the inactive filler does not practically affect the thermomechanical properties of PVC. Therefore, fine silica promotes a structuring of the PVC-matrix and the change in the internal energy plays a decisive role in reinforcement of the glassy PVC.

5.5 Biopolymers

Using the deformation calorimetry, Ebert et al. [194–196] studied the change in the internal energy of a number of solid poly-α-aminoacids, such as poly-L-alanin and its copolymers with S-cystein, poly-L-lysin and poly-γ-methyl-L-glutamate. Most of these studies were devoted to the determination of irreversible changes of the internal energy as a result of conformational changes accompanying irreversible deformation of these poly-α-aminoacids. Up to 1% stretching, a reversible energy-elastic behaviour is predominant: stretching of the films and fibres of poly-α-aminoacids is accompanied by endo-effects and contraction by exo-effects. The linear thermal expansion coefficients obtained from calorimetric measurements are in a good agreement with the data obtained by linear dilatometry. One can conclude that the general thermomechanical behaviour of solid poly-α-aminoacids is quite consistent with the thermomechanics of other solid polymers.

Godovsky et al. [197)] investigated the thermomechanics of the water-gelatin system. Gelatin which is obtained from nonenzymatic denaturation of collagen can be considered as a copolymer. The thermomechanical behaviour of water-gelatin systems strongly depends on the amount of water. If the water content is less than 25%, the samples exhibit at room temperature a typical glasslike behaviour. They are reversibly strained only by 1–2% and exhibit heat absorbtion. At ~25% of sorbed water, glass-to-rubber transition occurs. The internal energy of films containing 50–72% water decreases on stretching, which is connected with a negative energy contribution in the rubberlike water-gelatin systems. The thermomechanical heat inversion was not observed in these systems.

6 Concluding Remarks

The thermomechanical data accumulated during the last years as a consequence of the improvement of the technique of deformation calorimetry contributed significantly to the understanding of thermodynamics and mechanisms of the reversible deformation of polymers in the glassy, semicrystalline and rubbery state.

Two lines have been followed in the thermomechanical studies of polymers — along the phenomenological thermodynamics and along the thermomechanical equations of state. The second line allows to explain the molecular origin of the entropy and internal energy changes under deformation. It has been demonstrated in this review that during energy exchange deformation of solid and rubberlike polymers can always be described by a simple equation of state.

For the majority of polymer networks, the free energy of deformation at simple extension, compression, and torsion contains a significant intrachain energy contribution. The sign and value of the contribution is consistent with the isomeric state theory of chain conformations. This intrachain energy contribution is independent of the extent of deformation and agrees well with the prediction of the simple Gaussian theory of rubber elasticity. The temperature coefficients $d \ln \langle r^2 \rangle_0 / dT$ of the unperturbed chain dimensions obtained from the energy contribution $(\Delta U/W)_{V,T}$ are consistent with the data obtained by thermoelastic measurements and from the viscosity-temperature measurements in dilute solutions.

In contrast to intrachain changes of entropy and internal energy, which follow from the statistical theory, interchain changes of the internal energy, vibrational entropy and volume can be predicted by the statistical theory only at small strains ($\lambda < 1.3$).

In rubberlike materials such as filled elastomers, block and graft copolymers, the energy effects may depend on the content of filler or hard phase. In these cases, resolving the total effects on intra- and interchain ones is much more difficult than in unfilled polymer networks. At a large content, the filler and hard phase play an important role in interchain energy changes of the rubbery matrix. Although both filled rubbers and thermoelastoplastics exhibit the stress softening effect, in filled rubbers it is connected mainly with slippage of polymer chains over the surface of the filler particles, and in thermoelastoplastics it is caused by breakage of the initial stress-supporting hard phase.

The calorimetry of reversible deformation glassy and crystalline polymers has revealed two very important facts: (a) Glassy and crystalline polymers both in the undrawn and drawn state behave thermomechanically at simple stretching and compression in full agreement with the classical thermomechanical theory. (b) the two-phase crystalline polymers above the glass transition temperature of their amorphous regions cannot be considered an analogue of filled rubbers in which crystallites act as multifunctional crosslinks and fine fillers. A close molecular linkage of crystallites and amorphous regions causes the intermolecular effects in the amorphous regions to be dominated by deformation. This conclusion has been verified by thermophysical and mechanical data on drawn and undrawn crystalline polymers.

Note at proof: Recently Gaylord et al. (Polymer, 25, 1577, 1984) have shown theoretically that elastic deformation of crystalline polymers is controlled by energetic interaction rather than by entropy.

7 List of Symbols and Abbreviations

a_0	radius of unstrained sample
a, a_1, a_2	parameters of van der Waals equation of state of a network
e	uniform (volume) strain
f	retractive force
f_s	entropy component of retractive force
f_u	energy component of retractive force
p	number of statistical segments
r	end-to-end distance of network chains
r_{max}	maximum extension of network chains
$\langle r^2 \rangle$	mean square end-to-end distance of network chains in undeformed sample
$\langle r^2 \rangle_0$	mean square end-to-end distance of unperturbed chains
$\langle r^2 \rangle_i$	mean square end-to-end distance of network chains in the reference state
w	degree of crystallinity
x_i	generalized coordinate
A_0	cross-sectional area of sample
C	heat capacity
C	constant in the Gaussian equation of state for-rubber elasticity
C_1, C_2	constants in the Mooney-Rivlin equation
$D, D_m, \bar{D}_m$	strain functions in van der Waals equation of state
E	modulus of elasticity (Young modulus)
E_c	conformational energy of chains
E'	modulus of elasticity of filled rubbers
E_f	internal energy changes in filled rubbers
$E_\parallel$	modulus of elasticity along the draw axis
$E_\perp$	modulus of elasticity perpendicular to the draw axis
E_{cr}	modulus of elasticity of crystalline lattice
E_{AB}	modulus of elasticity of amorphous intrafibrillar regions

E_{AI}	modulus of elasticity of amorphous interfibrillar regions
F	free energy
F^*	Free energy of deformation of solids
G	free enthalpy
G^*	free enthalpy of deformation of solids
G_0	shear modulus
H	enthalpy
H^*	enthalpy of deformation of solids
K	modulus of elasticity (volume)
L_0	length of undeformed samples
L	length of deformed samples
L_i	length of samples in the reference state
M	twisting couple
P	pressure
Q	heat
S	entropy
T	absolute temperature
V	volume
U	internal energy
U^*	internal energy of deformation of solids
W	mechanical work
X	portion of interfibrillar amorphous regions
α	volume thermal expansivity
α_*	degree of deformation
β	linear thermal expansivity
$\beta_\parallel$	linear thermal expansivity along the orientation axis
$\beta_\perp$	linear thermal expansivity perpendicular to orientation axis
β_{cr}	linear thermal expansivity of crystalline lattice
β_{AB}	linear thermal expansivity of intrafibrillar amorphous regions
β_{AI}	linear thermal expansivity of interfibrillar amorphous regions
β_φ	linear thermal expansivity under the angle φ
γ	$= (\partial \ln \langle r^2 \rangle_0 / \partial \ln V)_{T,L}$
γ_s	shear deformation
γ'	$= \gamma/2$
δ	solubility parameter
ε	strain (uniaxial)
η	heat to work ratio
$\varkappa$	volume compressibility
$\varkappa_L$	linear compressibility
λ	elongation (or compression) ratio
λ_Q	elongation corresponding to inversion of heat
λ_u	elongation corresponding to inversion of internal energy
λ_m	limiting elongation
λ_f	elongation corresponding to inversion of force
μ	Poisson's ratio
ν	number of elastically active network chains
ξ, ξ_i	generalized force

σ	stress
τ	shear stress
φ	angle with the draw direction
ω	internal energy to work ratio
θ	twisting angle
PE	polyethylene
LDPE	low density polyethylene
HDPE	high density polyethylene
PP	polypropylene
PA	polyamide
PS	polystyrene
PET	poly(ethylene terephthalate)
NR	natural rubber
PDMS	polydimethylsiloxane
EPR	ethylene-propylene rubber
SBR	styrene-butadiene rubber
PCR	polychloroprene rubber
NBR	nitrile-butadiene rubber
PBR	polybutadiene rubber

8 References

1. Müller, F. H.: Thermodynamics of deformation. Calorimetric investigation of deformation processes, in: Rheology, New York: Academic Press, Vol. 5, 417 (1969)
2. Landau, L. D., Lifshiz, E. M.: Statistical Physics (in Russian), Moscow: Nauka 1976
3. Rummer, Yu. B., Ryvkin, M. Sh.: Thermodynamics, Statistical Physics and Kinetics (in Russian), Moscow: Nauka 1977
4. Callen, H. B.: Thermodynamics, New York: Wiley International 1960
5. Reiss, H.: Methods of Thermodynamics, New York: Blaisdell Publ. Co. 1965
6. Sharda, S. C., Tschoegl, N. W.: Macromolecules 7, 882 (1974)
7. Godovsky, Yu. K.: Thermal Physics of Polymers (in Russian), Moscow: Khimiya 1982
8. Godovsky, Yu. K.: Colloid Polym. Science 260, 461 (1982)
9. Kilian, H.-G.: Colloid Polym. Science 260, 895 (1982)
10. Nadai, A.: Theory of Flow and Fracture of Solids, New York: McGraw-Hill 1963
11. Wunderlich, B., Gaur, U.: Pure Appl. Chem. 52, 445 (1980)
12. Lifshiz, I. M.: J. Exp. Theor. Phys. 22, 475 (1952)
13. Kan, K. N.: Theoretical Questions of Thermal Expansion of Polymers (in Russian), Leningrad: Leningrad University Press 1975
14. Chen, F. C., Choy, C. L., Young, K.: J. Polym. Sci., Polym. Phys. Ed. 18, 2312 (1980)
15. Chen, F. C. et al.: J. Polym. Sci. Polym. Phys. Ed. 19, 971 (1981)
16. Wakelin, J. H., Sutherland, A., Beck, L. R.: J. Polym. Sci. 42, 278 (1960)
17. Choy, C. L., Chen, F. C., Young, K.: J. Polym. Sci., Polym. Phys. Ed. 19, 335 (1981)
18. Dadobaev, G., Slutsker, A. I.: Vysokomol. Soedin. A24, 1616 (1982)
19. Flory, P. J.: Trans. Farad. Soc. 57, 829 (1961)
20. Mark, J. E.: Rubber Chem. Technol. 46, 593 (1973)
21. Mark, J. E.: Rubber Chem. Technol. 55, 1123 (1982)
22. Allen, G. et al.: Trans. Farad. Soc. 67, 1278 (1971)
23. Price, C.: Proc. Roy. Soc. Ser. A 351, 331 (1976)
24. Godovsky, Yu. K.: Polymer 22, 75 (1981)
25. Gee, G.: Trans. Farad. Soc. 42, 585 (1946)
26. Guth, E., James, H. M., Mark, H.: Adv. Colloid Sci. 2, 253 (1946)

27. Shen, M.: Macromolecules 2, 358 (1969)
28. Tobolsky, A. V., Shen, M.: J. Appl. Phys. 37, 1952 (1966)
29. Shen, M., Kroucher, M.: J. Macromol. Sci. C12, 287 (1975)
30. Bleha, T.: Polymer 22, 1314 (1981)
31. Schwarz, J.: Polym. Bull. 5, 151 (1981)
32. Treloar, L. R. G.: Polymer 10, 291 (1969)
33. Treloar, L. R. G.: The physics of rubber elasticity, Oxford: Clarendon Press 1975³; in: Applied Fibre Science. Happey, F. (ed.), p. 103, London: Academic Press 1979
34. Dušek, K., Prins, W.: Strukture and elasticity of non-crystalline polymer networks, in: Adv. Polym. Sci. Berlin, Heidelberg, New York: Springer, Vol. 6, pp. 1–102 (1969)
35. Schwarz, J.: Habilitationsschrift, Clausthal Technical University 1978
36. Schwarz, J.: IUPAC Internat. Symp. Macromolecules, Mainz 1979, pp. 1370–1373
37. Schwarz, J.: Europhysical Conference on Macromolecules, Jablonna 1979, pp. 121–122
38. Mark, J. E., Curro, J. G.: J. Chem. Phys. 79, 5705 (1983); 80, 4521, 5262 (1984)
39. Shen, M.: J. Appl. Phys. 41, 4351 (1970)
40. Valanis, K. C., Landel, R. F.: J. Appl. Phys. 38, 2997 (1967)
41. Blatz, P. J., Sharda, S. C., Tschoegl, N. W.: Trans. Soc. Rheol. 18, 145 (1974)
42. Sharda, S. C., Tschoegl, N. W.: Macromolecules 9, 910 (1976)
43. Chang, W. V., Bloch, R., Tschoegl, N. W.: Macromolecules 9, 917 (1976)
44. Tschoegl, N. W.: Polymer 20, 1365 (1979)
45. Tschoegl, N. W.: IUPAC Internat. Symp. Macromolecules, Amherst 1982, pp. 862–
46. Ogden, R. W.: Proc. Roy. Soc. Ser. A 326, 565 (1972)
47. Chadwick, P.: Phil. Trans. Roy. Soc. London Sec. A 276, 371 (1974)
48. Ogden, R. W.: J. Mech. Phys. Solids 26, 37 (1978)
49. Gee, G.: Macromolecules 13, 705 (1980)
50. Kilian, H.-G.: Colloid Polym. Sci. 258, 489 (1980); 259, 1084 (1981)
51. Eisele, U., Heise, B., Kilian, H.-G., Pietralla, M.: Angew. Makromol. Chem. 100, 67 (1981)
52. Vilgis, Th., Kilian, H.-G.: Polymer 24, 949 (1983)
53. Flory, P. J.: Proc. Roy. Soc., London A351, 351 (1976)
54. Flory, P. J.: J. Chem. Phys. 66, 5720 (1977)
55. Flory, P. J., Erman, B.: Macromolecules 15, 800; 806 (1982)
56. Marrucci, G.: Rheol. Acta 18, 193 (1979)
57. Ball, R. C., Doi, M., Edwards, S. F., Warner, M.: Polymer 22, 1010 (1981)
58. Marrucci, G.: Macromolecules 14, 434 (1981)
59. Gaylord, R. J.: Polym. Bull. 8, 325 (1982); 9, 181 (1983)
60. Priss, L. S.: Pure Appl. Chem. 53, 1581 (1981)
61. Edwards, S. F.: Brit. Polym. J. 9, 140 (1977)
62. Greassly, W. W.: Adv. Polym. Sci. 47, 67 (1982)
63. Gottlieb, M., Gaylord, R. J.: Polymer 24, 1644 (1983); Macromolecules 17, 2024 (1984)
64. Godovsky, Yu. K.: Thermophysical Methods of Polymers Characterization (in Russian), Moscow: Khimiya 1976
65. Bloomquist, D. D., Sheffield, S. A.: J. Appl. Phys. 51, 5260 (1980)
66. Rodriquez, E. L., Filisko, F. E.: J. Appl. Phys. 53, 6536 (1982)
67. Wolf, F. P., Allen, G.: Polymer 16, 209 (1975)
68. Boyce, P. H., Treloar, L. R. G.: Polymer 11, 21 (1970)
69. Gent, A. N., Kuan, T. H.: J. Polym. Sci. 11, 1723 (1973)
70. Duvdevani, I. J., Biesnberger, J. A., Gogos, C. G.: Polym. Eng. Sci. 9, 250 (1969)
71. Godovsky, Yu. K., Slonimsky, G. L., Alekseev, V. F.: Vysokomol. Soedin. A11, 1181 (1969)
72. Molchanov, Yu. M., Molchanova, G. A.: Mechan. Polym. N5, 579 (1970)
73. Sergeev, Yu. A., Fainberg, E. Z., Michailov, N. V.: Vysokomol. Soedin. A14, 250 (1972)
74. Jehlar, P., Romanov, A., Pollak, V.: IUPAC Internat. Symp. Macromolecules, Mainz 1979, pp. 1409–1413
75. Andrianova, G. P., Arutunov, B. A., Popov, Yu. V.: J. Polym. Sci. Polym. Phys. Ed. 16, 1139 (1978)
76. Price, C., Evans, K. A., deCandia, F.: Polymer 14, 339 (1973)
77. Göritz, D., Müller, F. H.: Kolloid Z. Z. Polym. 252, 862 (1974)

78. DeCandia, F. et al.: J. Polym. Sci., Polym. Phys. Ed. *20*, 1525 (1982)
79. Allen, G., Price, C., Yoshimura, N.: Trans. Farad. Soc. *71*, 548 (1975)
80. Price, C., Allen, G., Yoshimura, N.: Polymer *16*, 261 (1975)
81. Chu, B.-T., in: A Critical Review of Thermodynamics. Stuart, E. B., Gal-Or, B., Brainard, A. J. (eds.), pp. 299–343, Baltimore: Mono Book Corp. 1970
82. Göritz, D., Müller, F. H.: Kolloid Z. Z. Polym. *251*, 679 (1973)
83. Mark, J. E., Flory, P. J.: J. Am. Chem. Soc. *86*, 138 (1964)
84. Mark, J. E.: J. Polym. Sci. Macromol. Revs. *11*, 135 (1976)
85. Godovsky, Yu. K.: Vysokomol. Soedin. *A19*, 2359 (1977)
86. Birshtein, T. M., Ptitsyn, O. B.: Conformations of Macromolecules (in Russian) Moscow: Nauka 1964
87. Boyer, R. F., Miller, R. L.: Internat. Conference on Rubber, Rubber-78, Kiev 1978, Preprint A_3.
88. Kilian, H.-G.: IUPAC Internat. Symp. Macromolecules, Amherst 1982, p. 566
89. Shen, M., Blatz, P. J.: J. Appl. Phys. *39*, 4937 (1968)
90. Treloar, L. R. G.: Polymer *19*, 1414 (1978)
91. Christensen, R. C., Hoeve, C. A. J.: J. Polym. Sci. *A-1,8*, 1503 (1970)
92. Mark, J. E.: Polym. Eng. Sci. *19*, 254 (1979)
93. Mark, J. E., Kato, M., Ko, J. H.: J. Polym. Sci. *C54*, 217 (1976)
94. Su, T.-K., Mark, J. E.: Macromolecules *10*, 120 (1977)
95. Kato, M., Mark, J. E.: Rubber Chem. Technol. *49*, 85 (1976)
96. Göritz, D., Müller, F. H.: Kolloid Z. Z. Polym. *251*, 679; 892 (1973)
97. Furukawa, J. et al.: Polym. Bull. *6*, 381 (1982)
98. Llorente, M. A., Mark, J. E.: J. Chem. Phys. *71*, 682 (1979)
99. Llorente, M. A., Mark, J. E.: J. Polym. Sci., Polym. Phys. Ed. *18*, 181 (1980)
100. Andrady, A. L., Llorente, M. A., Mark, J. E.: J. Chem. Phys. *72*, 2282; *73*, 1439; *74*, 2289 (1980)
101. Mark, J. E.: The use of model polymer networks to elucidate molecular aspects of rubber-like elasticity, in: Adv. Polym. Sci. Berlin, Heidelberg, New York: Springer Verlag, Vol. 44, pp. 1–26 (1982)
102. Zhang, Z. M., Mark, J. E.: J. Polym. Sci., Polym. Phys. Ed. *20*, 473 (1982); see also: Zhang, Z. M., Mark, J. E.: IUPAC Internat. Symp. Macromolecules Strasburg 1981, p. 725
103. Kilian, H.-G.: Colloid Polym. Sci. *259*, 1151 (1981)
104. Plate, N. A., Shibaev, V. P.: Comb-Like Polymers and Liquid Crystals (in Russian), Moscow: Khimiya 1980
105. Adv. Polym. Sci. Berlin, Heidelberg, New York: Springer Verlag, Vols. 59, 60, 61 (1984)
106. Jarry, J. P., Monnerie, L.: Macromolecules *12*, 316 (1979)
107. Rusakov, V. V., in: Mathem. Methods for Polymer Investigation pp. 54–55, Puschino 1983
108. Rusakov, V. V., in: Structural and Mechanical Properties of Composite Materials, pp. 37–48, Sverdlovsk, Sci. Papers of Ural Sci. Center of Akad. Nauk USSR 1984
109. Kock, H. J. et al.: Amer. Chem. Soc. Polym. Prepr. *24*, N2, 300 (1983)
110. Noshay, A., McGrath, J. E.: Block Copolymers. Overview and Critical Survey, New York: Academic Press 1977
111. Polmanteer, K. E., Lentz, C. W.: Rubber Chem. Technol. *48*, 795 (1975)
112. Kraus, G.: Rubber Chem. Technol. *51*, 297 (1978)
113. Warrick, E. L. et al.: Rubber Chem. Technol. *52*, 437 (1979)
114. Boonstra, B. B.: Polymer *20*, 691 (1979)
115. Rigbi, Z., in: Adv. Polym. Sci. Berlin, Heidelberg, New York: Springer Verlag, Vol. 36, 21 (1980)
116. Dannenberg, E. M.: Rubber Chem. Technol. *48*, 410 (1975)
117. Holden, G., Bishop, E. T., Legge, N. R.: J. Polym. Sci. *C26*, 37 (1969)
118. Tarasov, S. G., Godovsky, Yu. K.: Vysokomol. Soedin. *A22*, 1879 (1980)
119. Godovsky, Yu. K.: Makromol. Chem. Suppl. 6, 117 (1984)
120. Kawai, H., Hashimoto, T., in: Contemporary Topics in Polym. Science *3*, pp. 245–266, New York: Plenum Press 1979
121. Pedemonto, E. et al.: Polymer *16*, 531 (1975)

122. Papkov, V. S. et al.: Mechan. polym. N3, 387 (1975)
123. Godovsky, Yu. K., Tarasov, S. G.: Vysokomol. Soedin. *A19*, 2097 (1977)
124. Leblanck, J. L.: J. Appl. Polym. Sci. *21*, 2419 (1977)
125. Zapp, R. L., Guth, E.: Ind. Eng. Chem. *43*, 430 (1951)
126. Oono, R., Ikeda, H., Todani, I.: Angew. Makromol. Chem. *46*, 47 (1971)
127. Galanti, A. V., Sperling, L.: Polym. Eng. Sci. *10*, 177 (1970)
128. Tarasov, S. G. et al.: Vysokomol. Soedin. *A26*, 1077 (1984)
129. Pollak, V., Romanov, A.: Collect. Czech. Chem. Commun. *45*, 2315 (1980)
130. Romanov, A., Marcincin, K., Jehalr, P.: Acta Polym. *33*, 218 (1982)
131. Godovsky, Yu. K., Bessonova, N. P., Guseev, V. V.: Mechan. Polym. N4, 605 (1983)
132. Guseev, V. V., Shkalenko, Zh. I., Malinsky, Yu. M.: Vysokomol. Soedin. *A23*, 161 (1981)
133. Godovsky, Yu. K., Tarasov, S. G., Tsvankin, D. Ya.: IUPAC Internat. Symp. Macromolecules. Florence 1980, Vol. 3, pp. 235–238
134. Godovsky, Yu. K., Tarasov, S. G.: Vysokomol. Soedin. *A22*, 1613 (1980)
135. Leonard, W. J., Jr.: J. Polym. Sci. *C54*, 237 (1976)
136. Bonart, R., Morbitzer, L.: Kolloid Z. Z. Polym. *232*, 764 (1969)
137. Bonart, R.: Polymer *20*, 1389 (1979)
138. Hoffman, K., Bonart, R.: Makromol. Chem. *184*, 1529 (1983)
139. Godovsky, Yu. K., Bessonova, N. P., Mironova, N. N.: Vysokomol. Soedin. *A25*, 296 (1983)
140. Godovsky, Yu. K., Bessonova, N. P.: Colloid Polym. Sci. *261*, 645 (1983)
141. Godovsky, Yu. K., Bessonova, N. P.: Vysokomol. Soedin. *A19*, 2731 (1977)
142. Partridge, S. M., in: Chemistry and Molecular Biology of the Intercellular Matrix, Balazs, E. A. (ed.), Vol. 1, p. 593, London: Academic Press 1970
143. Weis-Fogh, T., Andersen, S. O., in: Chemistry and Molecular Biology of the Intercellular Matrix, Balazs, E. A. (ed.), Vol. 1, p. 671, London: Academic Press 1970
144. Weis-Fogh, T., Andersen, S. O.: Nature *227*, 718 (1970)
145. Gosline, J. M., Yew, F. F., Weis-Fogh, T.: Biopolym. *14*, 1811 (1975)
146. Grut, W., McCrum, N. G.: Nature *251*, 165 (1974)
147. Hoeve, C. A. J., Flory, P. J.: Biopolym. *13*, 677 (1974)
148. Andrady, A. L., Mark, J. E.: Biopolym. *19*, 849 (1980)
149. Dorrington, K. L., McCrum, N. G.: Biopolym. *16*, 1201 (1977)
150. Volpin, D., Cifferi, A.: Nature *225*, 382 (1970)
151. Evans, E. A., Hochmuth, R. M., in: Current Topics in Membranes and Transport, Bronner, F., Kleinzeller, A. (eds.), Vol. 10, p. 1, New York: Academic Press 1978
152. Evans, E. A., Skalak, R.: Mechanics and Thermodynamics of Biomembranes, Boca Raton, Florida: CRC Press, Inc., 1980
153. Evans, E. A., Waugh, R.: Colloid Interface Sci. *60*, 286 (1977)
154. Waugh, R., Evans, E. A.: Biophys. J. *26*, 115 (1979)
155. Gilmor, I. W., Trainor, A., Haward, R. N.: J. Polym. Sci., Polym. Phys. Ed. *16*, 1277 (1978)
156. Morbitzer, L., Hentze, G., Bonart, R.: Kolloid Z. Z. Polym. *216–217*, 137 (1967)
157. Godovsky, Yu. K., Bessonova, N. P., Konjuchova, E. V., Tarasov, S. G.: Paper presented at Rubber-84, Moscow 1984
158. Folkes, H. J., Keller, A., in: Physics of Glassy Polymers, Haward, R. N. (ed.), London: Applied Science Publishers 1973
159. Keller, A., Pedemonte, E., Willmouth, F. M.: Kolloid Z. Z. Polym. *238*, 385 (1970)
160. Folkes, H. J., Keller, A., Odell, J. A.: Am. Chem. Soc., Polym. Preprints *18*, 251 (1977)
161. Odell, J. A., Keller, A.: Polym. Eng. Sci. *17*, 544 (1977)
162. Krigbaum, W. R., Roe, R.-J., Smith, K. J., Jr.: Polymer *5*, 533 (1964)
163. Heise, B., Kilian, H.-G., Pietralla, M.: Progr. Colloid Polym. Sci. *62*, 16 (1977)
164. Lohse, D. J., Gaylord, R. J.: Polym. Eng. Sci. *18*, 512 (1978)
165. Butyagin, P. Yu., Garanin, V. V., Kuznetsov, A. R.: Vysokomol. Soedin. *A16*, 327 (1974)
166. Anisimov, S. P. et al.: Solid State Physics (FTT) *20*, 77 (1978)
167. Federov, Yu. N.: Vysokomol. Soedin. *B25*, 912 (1983)
168. Godovsky, Yu. K., Slonimsky, G. L., Papkov, V. S., Dikareva, T. A.: Mechan. polym. *5*, 785 (1970)
169. Thiele, J. L., Cohen, R. E.: Rubber Chem. Technol. *53*, 313 (1980)

170. Mead, W. T., Desper, C. R., Porter, R. S.: J. Polym. Sci., Polym. Phys. Ed. *17*, 859 (1979)
171. Capiati, N. J., Porter, R. S.: J. Polym. Sci., Polym. Phys. Ed. *15*, 1427 (1977)
172. Choy, C. L., Chen, F. C., Ong, E. L.: Polymer *20*, 1191 (1979)
173. Choy, C. L., Ito, M., Porter, R. S.: J. Polym. Sci., Polym. Phys. Ed. *21*, 1427 (1983)
174. Wolf, F.-P., Karl, V.-H.: Angew. Makromol. Chem. *92*, 89 (1980)
175. Wolf, F.-P., Karl, V.-H.: Colloid Polym. Sci. *259*, 29 (1981)
176. Wolf, F.-P., Karl, V.-H.: Makromol. Chem. *182*, 1787 (1981)
177. Marikhin, V. A., Myasnikova, L. P.: Supermolecular Structure of Polymers (in Russian), Leningrad: Khimiya 1977
178. Pakhomov, P. M. et al.: Vysokomol. Soedin. *A18*, 132 (1975)
179. Peterlin, A., in: Ultra-high Modulus Polymers, Ciferri, A., Ward, I. M. (eds.), p. 279, London: Applied Science Publishers 1979
180. Prevorsek, D. C. et al.: J. Macromol. Sci. Phys. *B9*, 733 (1974)
181. Chvalun, S. N. et al.: Vysokomol. Soedin. *B20*, 672 (1978)
182. Chvalun, S. N. et al.: Vysokomol. Soedin. *B22*, 359 (1980)
183. Chvalun, S. N. et al.: Vysokomol. Soedin. *A23*, 1381 (1981)
184. Chvalun, S. N. et al.: Internat. Symp. Chem. Fibres, Kalinin 1981, Vol. 1, pp. 116–121
185. Karpova, S. G. et al.: Vysokomol. Soedin. *A25*, 2435 (1983)
186. Marikhin, V. A., Myasnikova, L. P., Viktorova, N. L.: Vysokomol. Soedin. *A18*, 1302 (1976)
187. Godovsky, Yu. K. et al.: Internat. Symp. Chem. Fibres, Kalinin 1981, Vol. 1 (suppl.), pp. 66–72
188. Göritz, D., Müller, F. H.: Colloid Polym. Sci. *253*, 844 (1975)
189. Hellwege, K.-H., Hannig, J., Knappe, W.: Kolloid Z. Z. Polym. *188*, 121 (1963)
190. Wang, L.-H., Choy, C. L., Porter, R. S.: J. Polym. Sci., Polym. Phys. Ed. *20*, 633 (1982)
191. Kardos, J. L. et al.: Polym. Eng. Sci. *19*, 1000 (1979)
192. Shermergor, T. D.: Theory of Elasticity of Microheterogeneous Solids (in Russian), Moscow: Nauka 1977
193. Risyuk, B. D., Nosov, M. P.: Mechanical Anisotropy of Polymers (in Russian), Kiev: Naukova Dumka 1978
194. Ebert, G., Knispel, G., Müller, F. H.: Colloid Polym. Sci. *258*, 495 (1980)
195. Ebert, G. et al.: Progr. Colloid Polym. Sci. *67*, 175 (1980)
196. Ebert, G., Maeda, A., Müller, F. H.: Colloid Polym. Sci. *260*, 404 (1982)
197. Godovsky, Yu. K., Malzeva, I. I., Slonimsky, G. L.: Vysokomol. Soedin. *A13*, 2768 (1971)

Editor: K. Dušek
Received May 30, 1985

Carbon Black:
Surface Properties and Interactions with Elastomers

J. B. Donnet and A. Vidal
Centre de Recherches sur la Physico-Chimie des Surfaces Solides — CNRS
24, avenue Président Kennedy 68200 Mulhouse/France

We restrict, in this paper, the discussions related to the reinforcement of elastomers to the investigation of a single filler, carbon black. We, moreover, mostly focus on the part played by surface chemical interactions in the properties of filler reinforced rubbers.

After a screening of the different parameters available for the characterization of reinforcing fillers, the nature of filler-elastomer interactions is examined (occluded and bound rubber).

The effects exerted by the filler on the stress-strain properties and the modulus of the vulcanizates are investigated on the standpoint of the part played by the rubber immobilized on the filler and of the behavior of the elastomer in the vicinity of the filler.

In the last part of the paper, filler-elastomer chemical interactions which are able to take place through surface functional groups or surface reactive hydrogens are studied. The effect exerted by the created filler-elastomer bonds in the reinforcement process is then discussed.

1 Introduction 104

2 Characterization of Reinforcing Fillers 106

3 Nature of Filler-Elastomer Interactions 113
    3.1 Significance of Bound Rubber. 114
    3.2 Occluded Rubber 115

4 Effect of Filler-Elastomer Interactions on Vulcanizate Properties 117
    4.1 Low Strains. 117
    4.2 Moderate Strains 118
    4.3 High Strains 118
    4.4 Sample Failure 119

5 Filler-Elastomer Chemical Interactions 119

6 Conclusions. 126

7 References 126

1 Introduction

Natural and synthetic elastomers exhibit such a number of homogeneous physical and chemical properties that they form nowadays a large family able to meet a wide range of requirements. It follows that rubbers rank among the basic wares of the modern world. It is, however, obvious that their numerous domains of use result not only from their vulcanization but, above all, from their blending with fillers such as carbon black or silica.

Originally, these fillers were intended to play the role of extenders, in order to cut down manufacturing costs. Very soon, however, it was realized that some of them imparted unexpected properties (hardness, strength, ...) to the processed materials. This improvement of properties was called reinforcement.

The first reliable investigations related to the behavior of rubbers date back to the end of the 19th-beginning of the 20th century [1-4]. However, although much research work has been performed on this subject, one still may hope that a satisfactory explanation of the problems associated with reinforcement will be available by the year 2000 [5]. Anyhow, all investigators concerned with rubber reinforcement agree with this effect to be strongly dependent on an interfacial bonding between the surface of the filler particle and the elastomeric matrix.

Before proceeding any further, a more explicit definition of rubber reinforcement should be provided. First of all, this phenomenon is only related to the rubbery state of the elastomer. Fillers exhibit indeed little effect, if any, in the glassy state. Moreover, as a consequence of both the physical and chemical complexity of commercially significant elastomeric composites and the difficulties associated with the measurement of their mechanical properties, it follows that reinforcement is often defined through its effects [6]. Most important is the improvement of abrasion and tearing strength exhibited by carbon black-filled elastomers (Fig. 1) [7]. While noticeable in the case of natural rubber, these effects are of the utmost importance when considering synthetic elastomers, which appeared during the Second World War. Among all

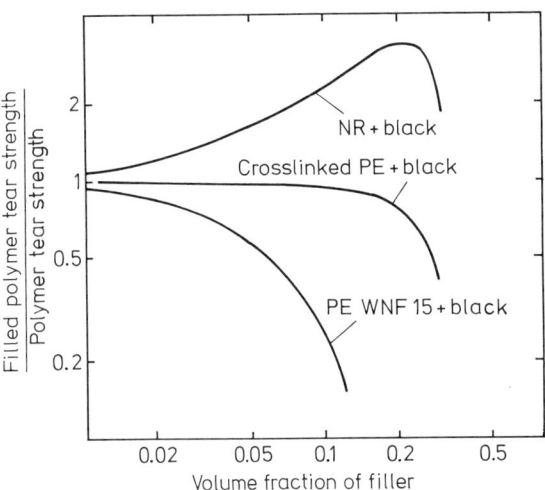

Fig. 1. Effect of carbon black on the tearing strength of a polymer [7]

fillers, only carbon black can efficiently improve their hardness and abrasion resistance, while the elastomers simultaneously retain their outstanding elastic properties. Until the appearance of synthetic elastomers, the only practical interest associated with carbon black-natural rubber blends was their enhanced abrasion and tearing resistance. It is nowadays universally acknowledged that reinforcement is associated with increased hardness, modulus and hysteretic phenomena, increased tensile, tear and fatigue strength, and hence with increased service life.

This phenomenon englobes, of course, many mechanisms [8]; Tables 1, 2, and 3 display the most significant ones. These may account for increased strength (Table 1), the increase of hysteretic phenomena (Table 2), and increased hardness or modulus (Table 3).

In this connection, Fig. 2 provides a qualitative illustration for interpreting modulus change of an elastomer upon filler blending [9]. A hydrodynamic or strain amplification effect, the existence of filler-elastomer bonds, and the structure of carbon black [10] all play a part in this modulus increase.

Table 1. Mechanisms contributing to increased strength [11]

1 — Strong or covalent interfacial bonds.
2 — Strain energy dissipation from increased hysteresis.
3 — Molecular alignment and stress distribution by surface slippage of network segments.
4 — Large aggregates distribute stress more equitably to attached polymer network segments.

Table 2. Mechanisms contributing to increased hysteresis [11]

A — Viscous contributions
   1 — Strain amplification of the rubber phase.
   2 — Molecular network segment surface slippage.
   3 — Aggregate displacement in network.
B — Bond breakage and reformation contributions
   4 — Interfacial surface bonds.
   5 — Interparticle molecular network segment bonds.
   6 — Secondary or transient filler structure bonds.

Table 3. Mechanisms contributing to increased stiffness, modulus, and rupture energy [11]

1 — Hydrodynamic viscosity increase
   a — Shape factor correction for effective volume concentration.
   b — Occluded volume correction.
2 — Increased stiffness and lower T_g of part of the rubber phase at interface due to surface interaction.
3 — Covalent bonding of network segments to filler surface.
4 — Range of surface bond strengths — weak to strong.
5 — Tightening of short chains between aggregates.
6 — Secondary or transient filler structure.
7 — Stress required for network segment slip rearrangements and alignment.
8 — Stress required for aggregate displacement in network.

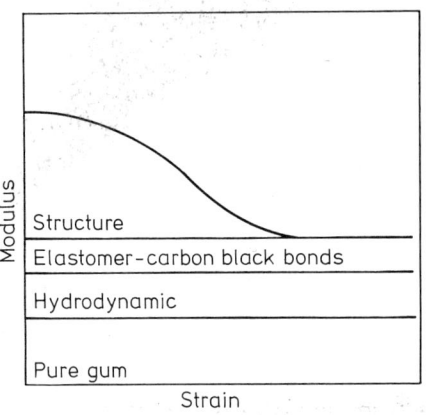

Fig. 2. Evolution with strain of different parameters playing a role in the modulus of carbon black reinforced elastomer (from Ref. [10])

It would be presumptuous to deal with such a tremendous field as elastomer reinforcement within the scope of a single review. Therefore, this discussion will be restricted to the investigation of a single filler, carbon black, and will mostly focus on the part played by surface chemical interactions in the properties of filler reinforced rubbers.

2 Characterization of Reinforcing Fillers

It is obvious that the morphological characteristics of the filler play a significant part in reinforcement. Indeed, large-sized particles dispersed in a gum only increase the stress applied to the elastomer, thus contributing to its weakening. Conversely, reinforcement effects are observed only when the particle diameter becomes sufficiently small, much lower than 10 µm in any case.

The single selection of particle diameter for the characterization of a reinforcing filler is, however, not appropriate, because, on the one hand, only fillers exhibiting a very poor reinforcing effect consist of independent spherical particles, and, on the other hand, gum-filler interactions taking place at the elastomer-filler interface are thus conditioned by the accessibility of the surface. The latter may, indeed, be restricted either by the presence of micropores or by the size of the macromolecule. The knowledge of the specific surface area of the filler is thus a prerequisite. Insofar as the determination of the filler specific surface area, performed by low-temperature gas adsorption or iodine adsorption, takes into account its microporosity, the adsorption of larger tensioactive molecules will often be favored [12,13].

The values of the particle-specific surface area and diameter are, beyond all questions, important elements for its characterization, but they are insufficient if taken alone. Figure 3, which displays the electron micrograph of a reinforcing black, shows indeed that carbon black, like any reinforcing filler, is not made of independent spherical particles, but displays aggregates consisting of more or less fused, roughly spherical nodules. The degree of particle aggregation, which is called "structure", depends of course on the number of nodules per aggregate. However, other parameters are also to be considered, e.g., its bulkiness, shape, or anisometry.

Thus, according to Medalia [14], the knowledge of four parameters is essential: the size of elementary particles making up the aggregate, the amount of solid material it contains, the radial distribution function of the solid material within the aggregate, and its anisometry. In short, it should be noted that the particle size d refers to the diameter of the quasispherical primary particles of which the aggregate is composed. The solid content or mass of an aggregate is a very basic property, which is nevertheless difficult to measure precisely. This quantity is obviously related to the number of particles N_p contained in the aggregate and to their diameter d. Using a technique of floc simulation, Medalia related N_p to the projected area A of the aggregate by the relation:

$$A = \pi \frac{d^2}{4} N_p^{0.87} .$$

This expression was subsequently confirmed by Sutherland et al. for flocs of up to 256 particles, using a corrected single-particle addition mechanism [15]. The radial distribution function is the concentration of solid matter per unit volume in successive shells proceeding outward from the center of mass of the aggregate. It results primarily from the determination of the quantity of liquid such as dibutyl phthalate (DBP) required for filling the voids of the aggregate under definite shear conditions. This technique, which can be applied either to an unconditioned carbon black or to a predensified sample (thus ruling out the effects of interaggregate agglomeration. which is called the secondary structure), is very effective in the case of carbon blacks, but much less satisfactory for other reinforcing fillers, such as graphitized carbon blacks or silicas. Finally, the anisometry allows for the irregular shape of the aggre-

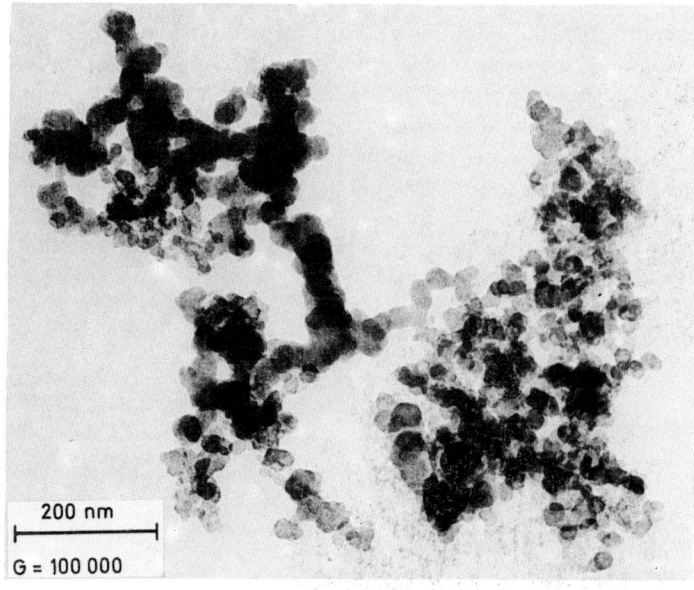

Fig. 3. Electron micrograph of a carbon black

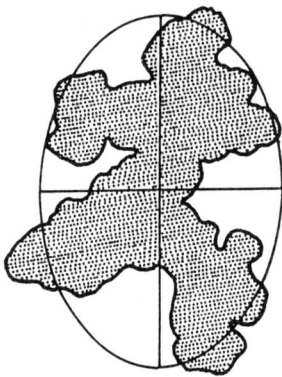

Fig. 4. Aggregate anisometry

gates. It is generally estimated by treating the projection of the electron micrograph like a two-dimensional figure (Fig. 4). The ratio of moments of inertia of silhouette about central principal axes provides the aggregate anisometry [14].

Some of the above-mentioned parameters are tabulated for a few carbon blacks of very different morphology (Table 4). It appears that, in spite of its complex morphology, the structure of a black can generally be fairly well evaluated knowing its specific surface area and its DBP absorption capacity (DBP number) [16].

From the technological standpoint, numerous empirical parameters have been used to characterize the effect of reinforcement. Two of them based on rheometry and filler-elastomer interactions will be discussed: the α_F parameter [17] and the reinforcement factor R_F [18-20].

The α_F Parameter:

Two parameters deduced from rheological properties (Monsanto rheometer) have been used to characterize the reinforcing ability of carbon black: the filler loading level and the value of the shear torque applied at the start of vulcanization D_a and at full cure D_∞, D_a^0, and D_∞^0, corresponding to the same values for the pure gum. If the curing mechanism of the elastomer is assumed to be unaffected by the filler, the increase of shear torque observed upon its blending can conceivably be attributed to the contribution to the overall crosslinking process of the polymer-filler

Table 4. Structural parameters for various furnace blacks [43]

ASTM code	d (nm)	DBP (cm³/100 g)	$A \times 10^2$ (µm²)	N_p	Anisometry	Surface area (m²/g)
HAF type						
N326	28	74	3.62	136	1.75	82
N330	29	101	7.22	278	1.78	83
N347	26	132	8.52	331	1.88	86
ISAF type						
N219	—	73	2.38	103	1.67	116
N220	22	116	5.45	261	1.82	112
N242	—	141	8.44	380	1.78	—

interactions. The latter can, therefore, be estimated and the α_F factor will be defined by the following relation:

$$\alpha_F = \frac{\left[\dfrac{D_\infty - D_a}{D_\infty^0 - D_a^0}\right] - 1}{m_f/m_p}$$

where m_f and m_p correspond, respectively, to the quantities of black and elastomer. From the relations

$$D_{spe} = \frac{D_\infty - D_a}{D_\infty^0 - D_a^0} \quad \text{and} \quad C_g = m_f/m_p$$

on arrives at the equation

$$D_{spe} = 1 + \alpha_F C_g$$

which is quite similar to that of Einstein. From α_F, one can directly estimate the part played by the morphological characteristics of the filler in the dispersion polymeric medium. The value of α_F of course depends on the nature of the polymer, and its change can be attributed to conformational peculiarities of the polymer chains in the vicinity of the surface. These conformations are, of course, in relation with the nature and extent of polymer-filler interactions. The value of α_F is thus able to provide valuable information as for the behavior of a filler blended with a given polymer (in situ characterization).

Another parameter, A, derived from ball-rebound values, has also been used in conjunction with α_F and surface area measurements to characterize carbon blacks in terms of vulcanizate properties:

$$R_0 - R = A m_f/m_p$$

where R_0 and R correspond, respectively, to the rebound of the unfilled and filled elastomer.

The reinforcement Factor R_F:

Looking for a test useful for the characterization of the fundamental properties of reinforcing fillers, Patel devised a method using a constant-force capillary rheometer to monitor the evolution of the viscoelastic properties of an elastomer upon addition of carbon black. The viscoelastic reinforcement factor R_F was defined as the ratio between the viscosity of the loaded and that of the unfilled rubber. Working with the same elastomer and various blacks used at different concentrations, Patel et al. showed that the R_F value is a function of the combined fundamental properties of the carbon black and the loading level. Each grade of black yields a typical R_F curve as a function of loading, such as those shown for N234, N326, and N762 in Fig. 5. The more reinforcing grades show a more rapid increase in R_F as the black loading increases. Comparisons at constant loading show a higher R_F for the blacks that provide the greatest reinforcement. It appears, moreover, (Figs. 6, 7, and 8) that good correlations

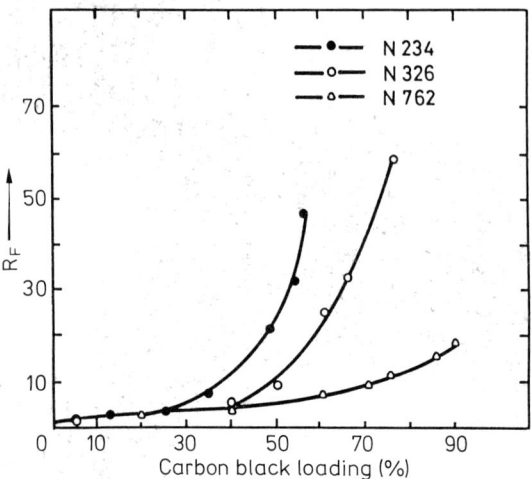

Fig. 5. Reinforcement factor versus loading [20]

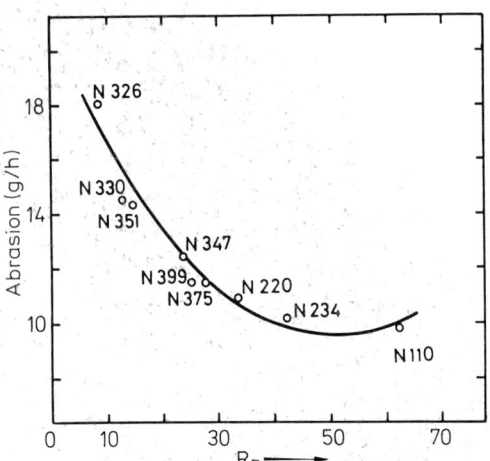

Fig. 6. Abrasion loss versus reinforcement factor: low oil SBR, 50 phr black [20]

exist between R_F and abrasion loss, flexometer running temperature, and pendulum rebound. Similarly, blacks of different structure were compared at constant R_F levels by studying the evolution of the 200% modulus of the composites versus the specific surface area (Fig. 9) or void volume (Fig. 10) of the filler. The plots show that the moduli separate in three categories and that in each of them they are independent of either the surface area or the structure of the filler. Identical results were obtained when studying the dependence of the 200% modulus on filler particle diameter. Thus, it appears that conventional blacks (furnace blacks) give higher moduli than the nonconventional ones (channel or acetylene blacks, ...) perhaps due to differences in surface activity depending on the nature of the reinforcing filler used. Finally, the study of the

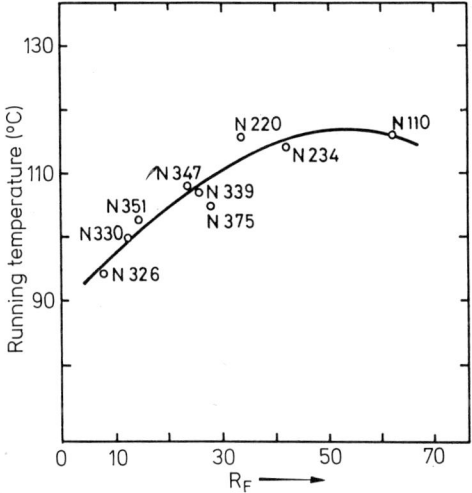

Fig. 7. Running temperature versus reinforcement factor: ASTM-NR, 50 phr black [20]

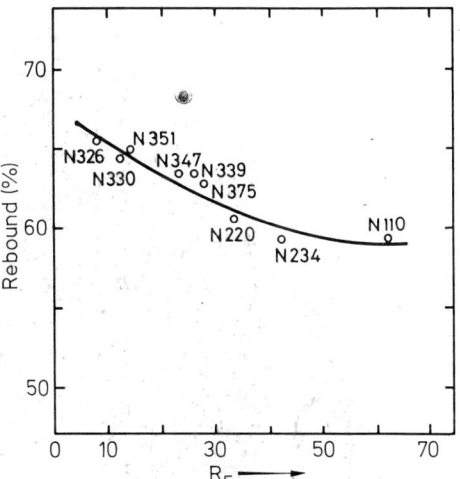

Fig. 8. Rebound versus reinforcement factor: ASTM-NR, 50 phr black [20]

dependence of the 200% modulus of a reinforced elastomer versus the size of its filler particles (Fig. 11) suggests a repartition of carbon blacks in two classes (particle diameters higher or smaller than 35 nm). Use of the reinforcement factor R_F allows thus the characterization of the properties of a carbon black by means of only one parameter.

Having defined the parameters allowing a characterization of the reinforcing ability of a black, what are the effects exerted by these aggregates on the properties of filled elastomers?

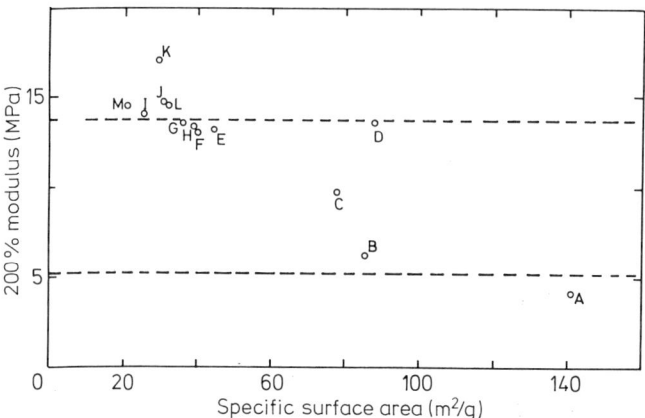

Fig. 9. 200% modulus of filled SBR 1500 versus carbon black specific surface area: loading (%) such as $R_F = 30$ [19)]

A = N110 (42.6%)
B = N326 (60.1%)
C = N351 (49.2%)
D = N358 (50.6%)
E = N550 (69.1%)
F = N650 (69.1%)
G = N630 (84.4%)

H = N660 (79.8%)
I = N762–25 (110.0%)
J = N762–30 (99.6%)
K = N774 (96.3%)
L = N787 (86.2%)
M = SL-90 (107.4%)

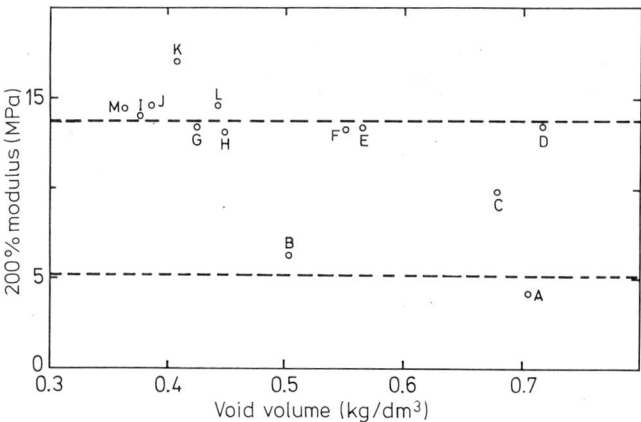

Fig. 10. 200% modulus of filled SBR 1500 versus carbon black void volume; loading (%) such as $R_F = 30$ [19)]

A = N110 (42.6%)
B = N326 (60.1%)
C = N351 (49.2%)
D = N358 (50.6%)
E = N550 (69.1%)
F = N650 (69.1%)
G = N630 (84.4%)

H = N660 (79.8%)
I = N762–25 (110.0%)
J = N762–30 (99.6%)
K = N774 (96.3%)
L = N787 (86.2%)
M = SL-90 (107.4%)

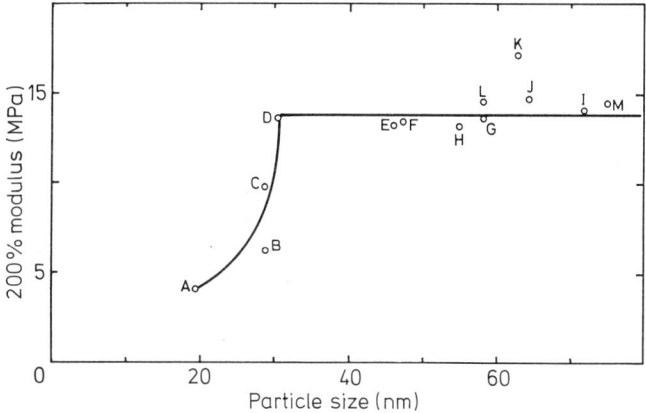

Fig. 11. 200% modulus of filled SBR 1500 versus carbon black particle size; loading (%) such as $R_F = 30$ [19)]

A = N110 (42.6%)
B = N326 (60.1%)
C = N351 (49.2%)
D = N358 (50.6%)
E = N550 (69.1%)
F = N650 (69.1%)
G = N630 (84.4%)

H = N660   (79.8%)
I = N762–25 (110.0%)
J = N762–30 (99.6%)
K = N774   (96.3%)
L = N787   (86.2%)
M = SL-90 (107.4%)

3 Nature of Filler-Elastomer Interactions

These interactions occur at the filler-elastomer interface which was shown by Hess and coworkers [21)] to be the place of adhesion forces acting between rubber and carbon black particles. These authors developed an electron microscopic technique which permitted a semi-quantitative evaluation of this adhesion process. Using carbon black-filled SBR, they were able to show that upon stretching a dewetting process of the solid particles was observed, particularly when the filler had a weak, if any reinforcing ability. The extensive cavitation which arises at the particle edges in such a case indicates the lack of adhesion of the corresponding black to the elastomer. Conversely, comparable tests performed with reinforcing blacks show a greatly reduced dewetting process, pointing at a much stronger particle-elastomer adhesion.

By plotting the percentage of carbon particles separated from the vulcanizate versus the stress applied to the sample during extension, Hess et al. determined the stress at which the arbitrary quantity of 20% of the black had been separated from the matrix. This stress was indicated as the adhesion index. It appears (Fig. 12) that blacks of higher structures are associated with an increase of the adhesion index, i.e., with an enhancement of filler-elastomer interactions.

The latter can be of two types, either purely mechanical, and be associated with the occlusion of rubber into carbon black aggregates (occluded rubber), are more complex and involve physical and chemical interactions, they will then be related to bound rubber.

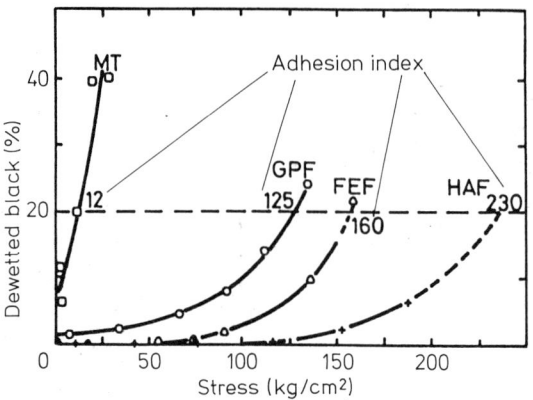

Fig. 12. Carbon separation as a function of stress in SBR [21]

3.1 Significance of Bound Rubber

When a uncured carbon black-elastomer blend is submitted for an extended period of time to a solvent extraction, only part of the elastomer can be recovered even with a very good solvent for the rubber. This quantity of unextractable elastomer adhering to the black is known as "bound rubber" or "carbon gel". An often held view is that bound rubber results from multicontact adsorption of elastomer chains. Desorption, which would then imply that all points of contact between the molecule and the surface be disengaged simultaneously, should be strongly dependent on the molecular weight of the elastomer. The data of Table 5 indicate that the amount of unextractable rubber is unaffected by the molecular weight of the polymer [22]. This points to the existence of bonding between particle and polymer at higher energy levels than to a mere physical adsorption and leads to the recognition of the role of chemisorption in the bonding of elastomers to carbon black.

A practical means of assessing the stability of the formed bond has been suggested by Sircar and Voet [22], who, for various carbon blacks and elastomers, plotted the unextractable amount of rubber from the filler as a function of the temperature of extraction for a variety or solvents. They observed the experimental points related to a given polymer-filler blend to be falling linearly, and the extrapolation to zero-grafting ratio provided the so-called solvolysis temperature T_m, which appears to be independent of the solvent used. T_m represents the temperature theoretically required to eliminate all bonds between carbon blacks and elastomer and is, therefore,

Table 5. Influence of molecular weight on unextractable elastomer [22]

Mol.wt. SBR (M_v)	$M_v/2000$	Unextracted (mg/g)	Ratio unextracted (%)
2000	1	45.7	1
13400	6.7	60.9	1.3
300000	150	145	3.2

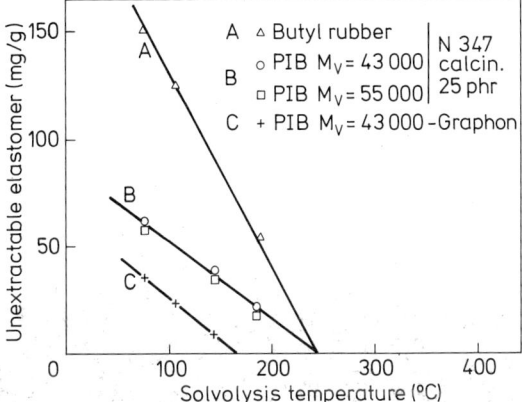

Fig. 13. Unextracted elastomer as a function of extraction temperature [22]

indicative of the bond strength. It appears (Fig. 13) that a graphitized carbon black, which exhibits only a very small amount of bound rubber, has a much lower bonding energy for the elastomer than a reinforcing black. This result confirms the major role played by chemisorption in the formation of bound rubber.

It is, however, necessary to discriminate between bound rubber and the effects it exerts on the neighboring elastomer molecules which form a rubber shell of finite thickness surrounding the carbon black particle. These effects associated with a reduced mobility of the rubber chains are affecting the physical properties of the sample. A study of the degree of immobilization of the rubber molecules, performed by nuclear magnetic resonance, showed the mobility of elastomeric chains adsorbed on carbon black to be extremely restricted within a thickness of about 0.5 nm [23]. Since chain mobility increases only gradually when the distance to the interface increases, the proximity of the surface is assumed within an interphase whose thickness is estimated at some 3 nm. These conclusions are in agreement with dilatometric measurements, which showed that, while the T_g of the elastomer close to the surface hardly changes (2 to 3 K), a broadening of the transition zone in the elastomeric domain was found to occur. This effect is to be ascribed to the above-mentioned rubber shell, in which the mobility of the polymer chains is very restricted [24]. Moreover, surface treatments can be expected to affect the size of the interphases and the mobility of their constitutive chains [25].

3.2 The Occluded Rubber

While bound rubber, resulting from elastomer chemisorption on the surface of the filler, was shown to modify ultimately the short-range mobility of the elastomeric chains surrounding the filler particle, the involvement of occluded rubber in the reinforcement process must be classified as purely mechanical (Fig. 14). The simple fact of occlusion is that the occluded elastomer is shielded to a significant extent from the deformation which the bulk of the elastomer undergoes when subjected to stress. As a consequence, occluded rubber is trapped by the carbon black aggregates without necessarily restricting the segmental motion of the chains located in the vicinity of

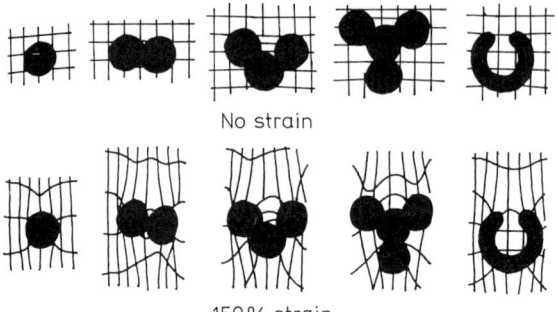

Fig. 14. Effect of strain on rubber in the vicinity of bidimensional models [14]

the filler. This mobility can, however, be affected if one bears in mind that entanglements may be trapped in occluded rubber [26]. Nevertheless, the concept of molecular occlusion leads to an increase of the effective volume of carbon black in rubber due to the filling of interparticulate voids by the elastomer. This can be interpreted, according to Kraus [27] and Medalia [28], by a structure-concentration equivalence principle, i.e., a high-structure carbon black used at low concentrations exerts the same effects as a low-structure carbon black used at high concentrations, since the rubber occluded in the high-structure carbon black acts as part of the filler rather than the matrix. This is exemplified in Fig. 15 for an SBR rubber reinforced with various amounts of furnace blacks of identical specific surface area but exhibiting widely different structures. It appears that the stress at 300% extension is a function of the effective carbon black concentration aV_2; V_2 and a being, respectively, the actual volume fraction of carbon black in the vulcanizate and a function of the black structure and vary linearly with the DBP number. Most of the properties of filled elastomers depend, however, on the specific surface area of the filler and, therefore, a simple structure-concentration relationship is valid only if this surface is known.

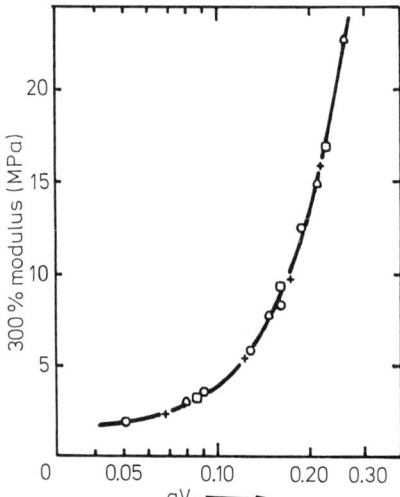

Fig. 15. 300% modulus versus effective carbon black concentration [3, 27]

To overcome this difficulty, Kraus and Janzen [16] suggested a correction by a surface factor. Thus, for a property P of a composite, a would be given by the relation:

$$P = P(aV_2)$$
$$a = (1 + C_1 \log S)(1 + C_2[D - 0.27])$$

in which V_2 has been defined previously, S is the accessible surface area (CTAB adsorption), and D the DBP number. When S is constant, a becomes a linear function of D. The empirical constants C_1 and C_2 allow for the relative sensitivity of P to surface and structure variation of the black. These equations were tested successfully for a number of properties, such as the Mooney viscosity, 300% modulus, etc., and are in agreement with various observations by Wolff and Westlinning [29, 30].

It was shown, on the one hand, that gum-filler interactions are associated with the immobilization of a certain amount of rubber on the surface or inside the carbon black aggregates, and, on the other hand, that the corresponding bound or occluded rubbers play important roles in the reinforcement process due either to a restriction of elastomer chain mobility in the vicinity of the filler or to an increase of the effective volume of the latter. What are now the effects exerted by a filler on the stress-strain behavior and the modulus of cured rubbers?

4 Effect of Filler-Elastomer Interactions on Vulcanizate Properties

This part will deal mainly with the behavior of rubber molecules in the vicinity of the aggregates. The energetics and fracture mechanics will be disregarded [readers interested in these subjects are referred to a recent paper by Medalia [5]]. These effects will be considered, depending on whether the strain applied is low (deformations, compression, or elongation, below about 5%), moderate (from about 5% to 15–25% in compression or elongation), high (over about 200%, in elongation), or is likely to induce failure.

4.1 Low Strains

At low strains two effects are to be considered: one is due to the occluded rubber, the other to the association of aggregates into agglomerates connected in a continuous network [31, 32]. When the agglomerate network effect is negligible (i.e., measurements carried out at high temperature or at low loadings), it can be shown that about half of the occluded rubber is effectively immobilized. The effective volume of filler is calculated according to the Guth-Gold equation:

$$E = E_0(1 + 0.67fc + 1.62f^2c^2)$$

relating the Young modulus of the composite to that of the unfilled elastomer, c being the filler volume fraction and f the shape factor used in the estimation of its anisometry. An increase of the effective volume of the filler is thus associated with an increase of the modulus of the vulcanizate. At usual filler loadings and temperatures, the effects exerted by the secondary structure contribute to an increase of the elastic and static moduli. The latter are dependent on the particle size or the specific surface area of the black.

4.2 Moderate Strains

At ordinary loadings and temperatures and in the moderate strain region, the effects associated with the secondary structure of the black have largely disappeared. Under these conditions, the application of the Guth-Gold equation (still legitimate) shows about half of the occluded rubber to be effectively immobilized. Under these conditions, the carbon particle size accounts essentially for the effects exerted by the filler on the hysteresis and loss tangent. These effects are related mainly to the breakdown and reformation of the secondary structure of the black, to the molecular slippage of the chains on the filler aggregates, and to the yet unstrained occluded rubber.

4.3 High Strains

At high strains, the Guth-Gold equation becomes unrealistic and the non-Gaussian behavior of the polymer chains has to be considered. In order to describe the corresponding behavior of filled elastomers, Mullins and Tobin [33], as well as Blanchard and Parkinson [34] and Bueche [35] suggested a modification of the statistical theory of elasticity. For an unfilled elastomeric network, the stress (σ) can be expressed as a function of the elongation (λ) by the semi-empirical Mooney-Rivlin equation:

$$\sigma = (2C_1 + 2C_2\lambda^{-1})(\lambda - \lambda^{-2}).$$

In this relation, $2C_2$ provides a correction for departure of the polymeric network from ideality, which results from chain entanglements and from the restricted extensibility of the elastomer strands. For filled vulcanizates, this equation can still be applied if it can be assumed that the major function of the dispersed phase is to increase the effective strain of the rubber matrix. In other words, because of the rigidity of the filler, the strain locally applied to the matrix may be larger than the measured overall strain. Various strain amplification functions have been proposed. Mullins and Tobin [33], among others, suggested the use of the volume concentration factor of the Guth equation to estimate the effective strain λ' in the rubber matrix:

$$\lambda' = \lambda(1 + 2.5c + 14.1c^2)$$

It follows that this strain amplification effect will be more important if a high structure carbon black is used. In this case, indeed, the real volume concentration of the filler will be significantly increased by the amount of occluded rubber trapped in the aggregates. At strains high enough, the occluded rubber, though anchored or immobilized by localized bonds, could be deformed. The magnitude of this deformation, which may involve some slippage of non-localized or mobile bonds, is probably not large enough to seriously affect the contribution of occluded rubber to the stress or modulus.

4.4 Sample Failure

When the rubber sample is stretched beyond the stress-relieving capacity of the occluded rubber or the slippage of weakly bonded chains [6], the fully stretched elastomer molecules start to break or rather to detach from the surface of the filler particles. As a consequence, a dewetting phenomenon is observed. When the remaining chains are insufficient to withstand the increased stress, catastrophic failure takes place [36, 37]. High-structure carbon black and, as a consequence, fine particle size implies a large concentration of load-bearing chains and therefore, a high tensile strength. Similar results, achieved when the number of filler-elastomer bonds is increased by chemical activation, suggest that the catastrophic failure is initiated at the polymer-filler bonds. It is probably induced at the periphery of an aggregate where the rubber is unshielded and the strain amplification effect may be high [14].

It thus appears that the effect of the structure of the filler on the ultimate properties of the composite involves mutually compensating effects. Higher-structure carbon blacks are associated with higher strain amplification degrees which tend to shorten the elongation at break and, presumably, increase the number of chains which, for a given deformation, are stretched to their breaking point. Conversely, the higher the structure of the black, the higher the number of chains bound to the filler surface and connecting the aggregates, thus delaying the failure and allowing the attainment of higher tensile strength. Similarly, upon deformation, occluded rubber must contribute to stress equalization and relief of the chains connecting the aggregates. Thus, for a given filler loading, there should always be an optimum level of structure for a particular ultimate property.

5 Filler-Elastomer Chemical Interactions

The surface of carbon blacks is energetically very heterogeneous. It exhibits indeed not only surface chemical functions, but also free valencies, network imperfections, ... However, only a small fraction of a carbon black-accessible surface is associated with high-energy sites. Rivin et al. [38] showed, for example, that 5% of the surface of a conventional furnace black corresponds to reactive sites able to induce a chemisorption process, whereas the remaining 95% are available for dispersive interactions (adsorption, ...). The latter while participating in the polymer-filler adhesion is, however, not contributing significantly to reinforcement. For example, a graphitized carbon black displays this type of adsorption, but its reinforcing ability is much lower than that of non-graphitized blacks; which points at the secondary role played by dispersive forces in the reinforcement process. The effect of these dispersive forces results mainly in a modification of the viscoelastic properties of the vulcanizates, since a certain degree of order takes place at the black-elastomer interface upon adsorption.

Table 6. Elementary analysis of a carbon black

Particle diameter (nm)		24.6
Specific surf. area (m²/g)		120
Elementary analysis (%)	C	96
	O	3.2
	H	0.55
Surface chemical (µeq/g) functionalities	—COOH	50
	—OH	100

This order is associated with chain alignment or, as previously seen, with a reduced mobility of the network strands.

The chemical view of reinforcement suggests a chemical reaction mechanism to take place between the elastomer and the surface functional groups of the black, yielding ultimately polymer chains grafted on the surface of the filler by means of covalent carbon-to-carbon bonding. What are these surface chemical functions? As shown by elemental analysis, a carbon black not only contains 96% carbon, but also 3.2% oxygen and 0.55% hydrogen (Table 6). These heteroatoms are part of several reactive functional groups on the solid surface; these groups have been identified as phenolic, hydroxyl, quinone, carboxyl, lactone, and peroxide groups among others.

Early investigations trying to demonstrate the involvement of chemical interactions in the reinforcement process attempted to correlate the surface chemical properties of carbon blacks with their reinforcing ability. Thus, the Hess adhesion index [21] has been used to assess the effect of the black-surface oxygen functional groups on filler-elastomer interactions. Some significant results obtained with reinforced polybutadiene or butyl rubber are presented in Table 7. Heating of carbon black under an inert atmosphere at temperatures of up to 900 °C results in a significant reduction of the surface oxygen-group content; the reverse being, of course, observed upon oxidation.

It appears from the evolution of the adhesion index that a distinction has to be made between the interactions carbon blacks are able to have with unsaturated or with saturated (or near-to-saturated) elastomers. Thus, the adhesion index of butyl rubber is enhanced upon oxidation of the black, while the reverse is observed with polybutadiene [38]. The improvement of the reinforcing ability of carbon black upon oxidation, in the former case, has been interpreted by Gessler [40] as due to chemical interactions of butyl rubber with active functional groups on the solid surface. Gessler, relating the reinforcing characteristics of the oxidized carbon black for butyl rubber to the presence of carboxyl groups on the surface of the filler, postulated a cationic

Table 7. Adhesion indexes (kg/cm²)

Polymer	Initial	Carbon black		
		Heat treated	Oxidized	Graphitized
Polybutadiene	86	125	56	19
Butyl rubber	77	83	112	13

reaction between the double bonds of the elastomer (Lewis base) and the carboxyl groups of the black. Later, Voet [41] failed, however, to confirm this mechanism.

Upon graphitization, by heating in an inert atmosphere at temperatures above 2700 °C, all surface chemical groups as well as hydrogen and all high-energy adsorption sites are removed. The results (Table 7) indicate an important decrease of the adhesion index, independent of whether the polymer is saturated or not. The carbon black then has lost a significant part of its reinforcing characteristics.

It thus appears that the beneficial effects of surface oxygen groups are only felt in the case of butyl rubber, which is comparatively inert from the chemical standpoint. Such results do not mean, however, that the occurrence of other chemical interactions at the filler-elastomer interface have to be discarded. Indeed, while heat treatment of the black, in order to remove surface oxygen groups, does not affect butyl rubber reinforcement, the adhesion indexes of polybutadiene or SBR are strongly decreased (Table 7). Other chemical interactions between elastomer and carbon blacks can be involved such as, for example, those associated with the black's ability to behave as a free radical acceptor [42-44]. Macroradicals formed during the mastication of filler-elastomer blends can indeed be deactivated on the surface of the black, thus leading to grafted materials [45,46].

Beside the extensively investigated oxygen chemical groups, hydrogen atoms are located on the surface of the filler. These peripheral hydrogens are bonded to edge carbon atoms of the graphitic crystallites of which the black is composed. Their reactivity is dependent on their electronic environment, the proximity of surface chemical functions, and structural defects. The occurrence of more or less reactive hydrogens is thus conceivable. It is, however, impossible to relate the total hydrogen content of a black, e.g., as calculated from elemental analysis to its reinforcing power. Yet, if upon removal of the surface oxygen groups the hydrogen directly bonded to the carbon becomes accessible, it would be possible to assess its effect in elastomer reinforcement. While some investigators had already suggested a correlation between the properties of a vulcanizate and the hydrogen content of the black [47] or had attempted to set up a relation between the number of bonds able to be created per unit surface area and the quantity of hydrogen able to react with bromine [48], we had to wait until the end of the sixties [49,50] for showing clearly the part played by the carbon-black surface hydrogens in reinforcement.

The relationship of hydrogen content to reinforcement was investigated by answering the three following questions:
1. Are surface hydrogens of carbon black reactive with SBR or, in other words, is there any hydrogen transfer?
2. Is hydrogen reactivity related to reinforcement, or is there any relationship between hydrogen content and strain energy?
3. Is the total hydrogen of the black important or only a more reactive part of it?

By answering the first question, it was possible to point at the occurrence of hydrogen exchange between a tritium-labeled SBR and carbon black. Part of the surface hydrogen was, indeed, shown to transfer from the black to the polymer, and conversely, without necessarily affecting elastomer reinforcement. In order to check this point and examine whether the presence of hydrogen on the surface of carbon black is related to its reinforcing ability and whether the latter is influenced by the total hydrogen content, tritium-labeled carbon black was progressively heated at increas-

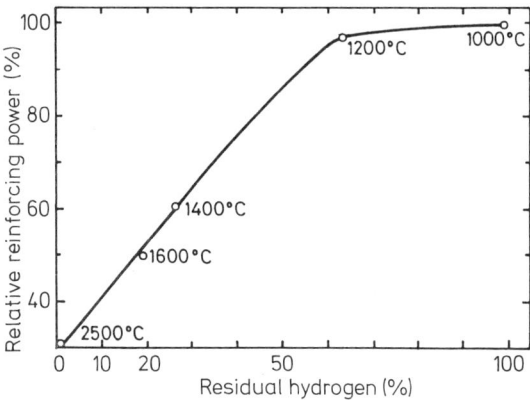

Fig. 16. Strain energy retained versus residual hydrogen ratio [68]

ing temperatures. The reinforcing ability of the black as derived from stress-strain curves was then compared with the residual surface hydrogen content. Figure 16 shows that a correlation is obtained between the latter and the strain energy retained by the vulcanizate.

However, heating carbon black at temperatures higher than 1400 °C not only removes the surface hydrogen atoms but also can be associated with structural modifications (even if graphitization is not yet important at such temperature). As a consequence, the decrease of the reinforcing potential of the black can result either from hydrogen removal or from a rearrangement followed by the possible elimination of structural defects. These results, nevertheless, suggest the possibility of a chemisorption mechanism involving hydrogen abstraction from either carbon black or rubber (or both) resulting in polymer grafting on the surface of the black. Serizawa et al. [25] have also demonstrated the role played by these surface hydrogen atoms in bound rubber formation in the case of natural rubber. They indeed showed, by NMR, that a decrease in the number of hydrogen atoms located at the edges of graphitic planes was associated with an enhanced mobility of the bound rubber chains. This evolution can be interpreted as resulting from an enlargement of the size of polymer loops bonded to the surface of the filler, connected with a decrease of the interactions carbon black is able to have with natural rubber.

To point to the existence, on the surface of the black, of hydrogen atoms differing in reactivities, the kinetics of the tritium-labeling reaction was investigated [50]. The results obtained for a series of carbon blacks exhibiting different reinforcing abilities are reported in Table 8, assuming a first-order substitution kinetics for all varieties of hydrogen. The quantity k′ stands for the apparent rate constants of the substitution reactions and x for the hydrogen relative proportion. Four types of surface hydrogen could thus be identified. It appears, moreover, that the labeling rate constants for a given type are all of the same order of magnitude, but differ significantly from one type to the other. The relative proportions of each type vary, however, depending on the nature of the black, the evolution of the most reactive hydrogens correlating quite well with the reinforcing potential of the filler. They could thus play a very significant part in the reinforcement of elastomers.

All these statements have been substantiated by studying the substitution of

Table 8. Hydrogen exchange reaction [50]

	Aro 1LS	Aro 3LS	Aromex	Aro 100	Aro 150	Aro 175
k'_1 (h^{-1})	0.090	0.097	0.080	0.106	0.101	0.105
x_1 (%)	2.0	2.7	5.3	7.9	9.3	10.2
k'_2 (h^{-1})	0.055	0.051	0.040	0.065	0.062	0.064
x_2 (%)	13.4	24.9	9.9	8.8	22.9	17.5
k'_3 (h^{-1})	0.028	0.025	0.020	0.030	0.026	0.026
x_3 (h^{-1})	40.6	31.7	25.3	29.5	24.6	37.9
k'_4 (h^{-1})	0.021	0.021	0.018	0.026	0.025	0.025
x_4 (h^{-1})	44.0	40.7	59.5	53.8	43.8	32.6

surface hydrogens by chlorine atoms. This reaction performed by using tritium-labeled carbons was investigated at various temperatures and was traced by loss of tritium. Figure 17 shows that tritium elimination starts at room temperature and that all of the surface hydrogen has reacted at 750 °C. It is, moreover, noteworthy that the elimination of hydrogen takes place in four steps, each one being probably associated with a peculiar type of reactive hydrogen. The first and second steps correspond to about 10% and 18% hydrogen loss, respectively. These proportions are in agreement with those of type 1 and 2 hydrogens yielded by kinetic experiments. These studies reveal, therefore, that the reinforcing potential of carbon black towards elastomers is not related to the total content of surface hydrogen atoms, but is directly dependent on the ratio of their most reactive form.

It appears, beyond all doubt, that filler-elastomer interactions result in the formation of chemical bonds between the polymer and the solid surface, which are due to a reaction of the macromolecule either with the surface chemical groups or with the surface hydrogen atoms. Is, however, the formation of covalent filler elastomer bonds a prerequisite for reinforcement to occur?

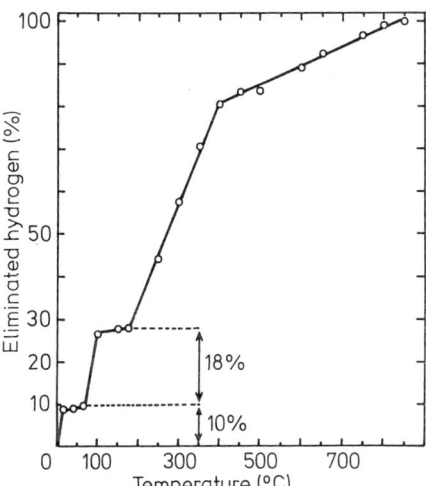

Fig. 17. Effect of chlorine treatment temperature on hydrogen elimination [69]

Table 9. Effects of carbon blacks graphitization on SBR reinforcement [52]

	Carbon black A		Carbon black B	
	Initial	Graphitized	Initial	Graphitized
Spec. surface (m^2/g)	116	86	108	88
DBP Abs. (cm^3/g)	1.72	1.78	1.33	1.54
Bound rubber (%)	34.4	5.6	30.6	5.8
300% modulus (MPa)	14.4	3.5	10.3	2.9
Tensile str. (MPa)	26.2	23.4	27.6	22.7
Elong. at break	4.50	7.30	6.30	7,50
Rel. abrasion res.	100	34	100	17

Tensile strenght of unfilled SBR: 1.4 MPa

Use of graphitized carbon blacks provided an answer to this question. Indeed, the graphitized blacks, while exhibiting the main features of carbon black morphology, are unquestionably the carbon forms closely related to a chemically inert filler. Graphitization of carbon black is associated, indeed, with a more regular crystallographic structure, while the morphology, the surface area, and the structure of the original carbon black remain unchanged [51]. It appears (Table 9) that graphitization does not induce substantial specific surface-area or structure modifications. It is, however, associated with an important decrease of the amount of bound rubber obtained upon blending a graphitized black with SBR [52]. The same applies to the properties of the corresponding vulcanizates; their 300% modulus and abrasion resistance are significantly lowered. Conversely, their tensile strengths appear to be practically unaffected.

Graphitized carbon blacks, thus undoubtly display reinforcing abilities which become obvious when considering the tensile strength of the unfilled vulcanizate. It follows that the formation of a filler-elastomer chemical bond is not a requirement for reinforcement to occur. It strongly participates, however, in its effectiveness, and determines the good mechanical properties connected with rubber reinforce-

Table 10. Effect of promoter[a] on the properties of vulcanized butyl rubber [54].

Properties	Promoter	
	0%	1%
Mooney viscosity	79	95
Bound rubber (%)	33.5	50
Tensile strength (kg/cm^2)	82	125
300% modulus (kg/cm^2)	62	118
Shore hardness A2	61	54
Electric resistivity (Ω cm)	10^4	10^{15}
Running temperature (°C)	17	14
Permanent set (%)	2.6	1.4

[a] bis(p-nitrosophenyl)-1,4-p&perazine

Table 11. Properties of vulcanizates filled with a polystyrene HAF grafted carbon black [64]

Polystyrene (%)	200% modulus (kg/cm²)	Tensile strength (kg/cm²)	Elongation at break (%)	Shore hardness
Cis-polybutadiene				
0.35	21	202	695	52
0.49	17	210	705	52
0.85	14	195	745	52
1.4	14	173	740	48
3.3	10	180	810	46
EPR				
0.35	28	271	605	55
0.49	26	268	605	55
0.85	21	258	680	53
1.4	18	261	705	52
3.3	10	228	785	48

ment. Improvements in the reinforcement process are indeed observed when promoters, such as N-4-dinitroso-N-methylaniline [53] or bis(p-nitrosophenyl)-1,4-piperazine [54], are used in order to generate carbon black-elastomer bonds (Table 10).

Another way for increasing filler-elastomer interactions could be the grafting of a polymer on the solid surface. A number of methods exist to secure the attachment of macromolecules to the surface of carbon black particles; e.g., a polymeric chain may be grown on an initiation site on the surface, small molecules previously attached to the surface may be copolymerized with a monomer, a polymeric chain, either radical, cationic, or anionic in nature, may be terminated on an active site of the solid surface, etc. [55-63].

It appears (Table 11) that the reinforcing ability exhibited by a HAF carbon black in polybutadiene or EPR is much lower after its surface has been previously modified by grafting of polystyrene [64]. Such results cannot, however, be associated with the elastomer's incompatibility towards the polystyrene grafts since polyisoprene-grafted carbon blacks yield similar conclusions [65].

Such results can be interpreted by considering, on the one hand, that the grafted polymer, immobilized on the surface, is thus prevented from exchanging any interaction with the elastomeric matrix, and, on the other hand, that an oxidative aging of the grafts could not be ruled out as a consequence of an exposition of the modified black to the atmosphere in the course of its preparation. By avoiding any possibility of oxidative process to occur, it was shown [66] for low strains and using dynamic-mechanical measurements, i.e., under conditions approximating those encountered in practice for filled elastomers, that the formation of filler-polymer chemical bonds exerted a positive effect either from the standpoint of energy dissipation in the interphase or that of an improved stress distribution (energetically more uniform surface).

6 Conclusions

1 — The reinforcement of elastomers is determined by the morphology of the filler and the chemical and physical interactions it is able to exchange with the polymer.
2 — The morphology of a filler, such as carbon black, is usually well estimated by its specific surface area accessible to the rubber molecules and by its density.
3 — Filler-elastomer interactions are of two types:
   — purely mechanical: occluded rubber, partly shielded from the stress, it contributes to an increase of the filler content.
   — due to physical adsorption of chemisorption: bound rubber modifies at finite distance the mobility of the elastomeric chains surrounding the filler particles.
4 — The effect exerted by the filler on the stress-strain properties and the modulus of the vulcanizate has been discussed from the standpoint of the part played by the rubber immobilized on the filler and of the behavior of the elastomer in the vicinity of the surface.
5 — The filler-elastomer chemical interactions take place through its surface functional groups and hydrogen atoms. Coupling agents improve polymer-filler adhesion. From the point of view of dynamic-mechanical properties for low strains, the filler-elastomer bonds have a positive effect in the reinforcement process.

7 References

1. Heinzerling, C., Pahl, W.: Verh. Ver. Behord. Gewerbefl. *1891*, 415; *1892*, 25
2. Mote, S. C.: Quotation from Stern, H. J., for the admission of Sydney Charles Mote in the International Rubber Science Hall of Fame, University of Akron, Akron Ohio, November 9, 1984
3. Ditmar, H.: Gummi Ztg. *20*, 394, 733, 844, 1077 (1906)
4. Wiegand, W. B.: IRI Trans. *1*, 141 (1925–1926)
5. Medalia, A. I.: International Conference on Structure-Property Relations of Rubber, Kharagpur India, December 1980
6. Dannenberg, E. M.: Rubber Chem. Techn. *48* (3), 411 (1975)
7. Kendall, K., Sherliker, F. R.: Brit. Polym. J. *12*, 85 (1980)
8. Kraus, G.: Rubber Chem. Techn. *51* (2), 297 (1978)
9. Payne, A. R.: "Dynamic properties of filler-loaded rubbers". In: Reinforcement of elastomers. Kraus, G. (Ed.). New York: Interscience 1965, pp. 69–123
10. Meinecke, E. A., Maksin, S.: Coll. Polym. Sci. *258*, 556 (1980)
11. Dannenberg, E. M.: "Influence of filler-elastomer interactions on reinforcement behavior". In: Le renforcement des élastomères. Paris: CNRS 1975, pp. 129–135
12. Lammond, T. G., Price, C. R.: Rubber Chem. Techn. *43*, 941 (1970)
13. Janzen, J., Kraus, G.: Rubber Chem. Tech. *44*, 1287 (1971)
14. Medalia, A. I.: "Filler aggregates and their effect on reinforcement". In: Le renforcement des élastomères. Paris: CNRS 1975, pp. 63–79
15. Sutherland, D. N., Goodarz-Nia, I.: Eng. Sci. *26*, 2071 (1971)
16. Kraus, G>, Janzen, J.: Kautsch. Gummi Kunstst. *28*, 253 (1975)
17. Wolf, S.: Kaut. Gummi. Kunstst. *23*, 7 (1970)
18. Patel, A. C.: Plast. Rubber Proc. *5*, 74 (1980); Carbon Blackboard *4* (6)
19. Patel, A. C., Byers, J. T.: Elastomerics *112* (8), 17 (1980); Ibid. *114* (2), 29 (1982)
20. Byers, J. T., Patel, A. C.: Rubber World *188* (3), 21 (1983)

21. Hess, W. M., Lyon, F., Burgess, K. A.: Kautsch. Gummi Kunstst. *20* (3), 135 (1967)
22. Sircar, A. K., Voet, A.: Rubber Chem. Tech. *43*, 973 (1970)
23. Kaufmann, S., Slichter, W. P., Davis, D. D.: J. Polym. Sci. A2, *9*, 829 (1971)
24. Kraus, G., Gruver, J. T.: J. Polym. Sci. A2, *8*, 571 (1970)
25. Serizawa, H., Nakamura, T., Ito, M., Tanaka, K., Nomura, A.: Polym. J. *15* (7), 543 (1983)
26. McArthur, A., Stepehens, H. L.: J. Appl. Poly. Sci. *28*, 1561 (1983)
27. Kraus, G.: Rubber Chem. Tech. *44*, 199 (1971); J. Appl. Polym. Sci. *15*, 1679 (1971)
28. Medalia, A. I.: J. Colloid Interface Sci. *32*, 115 (1970)
29. Westlinning, H., Wolff, S.: Kautsch. Gummi Kunstst. *19*, 470 (1966)
30. Wolff, S.: Kautsch. Gummi Kunstst. *27*, 511 (1974)
31. Payne, A. R.: J. Appl. Polym. Sci. *6*, 57 (1962)
32. Medalia, A. I.: Rubber World *168* (5), 43 (1974)
33. Mullins, L., Tobbins, N. R.: J. Appl. Polym. Sci. *9*, 2993 (1965)
34. Blanchard, A. F., Parkinson, D.: Second Rubber Technol. Conference, London, Great Britain, 1948
35. Bueche, F.: "Network theories of reinforcement". In: Reinforcement of elastomers. Kraus, G. (Ed.). New York: Interscience 1965, pp. 1–22
36. Bueche, F.: J. Polym. Sci. *24*, 189 (1957)
37. Buckler, E. J.: Plastics Rubber Intern. *4*, 255 (1979)
38. Rivin, D., Aron, J., Medalia, A. I.: Rubber Chem. Tech. *41*, 330 (1968)
39. Sweitzer, C. W., Burgess, K. A., Lyon, F.: Rubber World *143* (5), 73 (1961)
40. Gessler, A. M.: Rubber Chem. Tech. *42*, 850 (1969)
41. Voet, A.: Kautsch. Gummi Kunstst. *26*, 254 (1973)
42. Watson, M.: "Chemical Interaction of fillers and rubbers during cold milling". In: Reinforcement of elassomers. Kraus, G. (Ed.). New York: Interscience 1965, pp. 247–260
43. Donnet, J. B., Voet, A. (Eds.): Carbon Black. New York: Dekker 1976
44. Waldrup, M. A., Kraus, G.: Rubber Chem. Tech. *42*, 1155 (1969)
45. Jamroz, M., Kozlowski, K., Sieniakowski, H., Jachym, B.: J. Polym. Sci. *15*, 1359 (1977)
46. Cashell, E. M., McBrierty, V. J.: J. Mater. Sci. *12*, 2011 (1977)
47. Studebaker, M. L.: Rubber Chem. Tech. *30*, 1400 (1957)
48. Rehner, J.: "The nature of polymer-filler attachments". In: Reinforcement of elastomers. Kraus, G. (Ed.). New York: Interscience 1965, pp. 153–186
49. Papirer, E., Voet, A., Given, P. H.: Rubber Chem. Tech. *42* (4), 1200 (1969)
50. Papirer, E., Donnet, J. B., Heinkele, J.: J. Chim. Phys. *68*, 581 (1971)
51. Schaeffer, W. D., Smith, W. R., Polley, M. H.: Ind. Eng. Chem. *45*, 1721 (1953)
52. Brennan, J. L., Jermyn, T. E., Boonstra, B. B.: J. Appl. Polym. Sci. *2687* (1964)
53. Walker, L. A., Kenwood, J. E.: Rubber Age *90*, 925 (1965)
54. Fabre, R., Bertrand, G.: Rev. Gen. Caout. Plast. *42*, 405 (1965)
55. Donnet, J. B., Henrich, G.: J. Polym. Sci. *46*, 277 (1960)
56. Donnet, J. B., Henrich, G., Riess, G.: Rev. Gen. Caout. Plast. *39*, 583 (1962)
57. Donnet, J. B., Geldreich, L., Henrich, G., Riess, G.: Rev. Gen. Caout. Plast. *41*, 519 (1964)
58. Donnet, J. B., Peter, G., Riess, G.: J. Polym. Sci. *22*, 645 (1969)
59. Donnet, J. B., Vidal, A., Riess, G., Geldreich, L.: Rev. Gen. Caout. Plast. *47*, 1289 (1970)
60. Donnet, J. B., Vidal, A., Riess, G.: J. Chim. Phys. *68*, 1642 (1971)
61. Donnet, J. B., Riess, G., Majowski, G.: Eur. Polym. J. *7*, 1065 (1971)
62. Papirer, E., Donnet, J. B., Riess, G., Van Tao, N.: Angew. Makromol. Chem. *19*, 65 (1971)
63. Drappel, S., Gauthier, J. M., Franta, E.: Carbon *21* (3), 311 (1983)
64. British Patent 1,103,855 (1968) (to Chemische Werke Hüls) Germ. Appl. C33047 (1964)
65. Dannenberg, E. M., Papirer, E.: Rev. Gen. Caout. Plast. *51*, 823 (1974)
66. Lebras, J., Papirer, E.: J. Appl. Polym. Sci. *22*, 525 (1978)
67. Medalia, A. I.: J. Colloid Interface Sci. *24*, 393 (1967)
68. Voet, A., Papirer, E., Aboytes, P., Schultz, J.: Rev. Gen. Caout. Plast. *48* (9), 935 (1971)
69. Donnet, J. B.: Brit. Polym. J. *5*, 213 (1973)

Editor: K. Dušek
Received June 26, 1985

Functionality and Molecular Weight Distribution of Telechelic Polymers

S. G. Entelis, V. V. Evreinov and A. V. Gorshkov
Institute of Chemical Physics, USSR Academy of Sciences, Kosygina 4,
117334 Moscow/USSR

Telechelic polymers are widely used to obtain various types of polymers. Of exceptional importance in the behaviour of telechelic polymers is the nature and number of reactive functional groups (RFG). Along with molecular weight distribution (MWD), a telechelic polymer is to be characterized by a new quantity, introduced by us, — the functionality type distribution (FTD). This characteristic determines to a considerable extent the topology of the resulting polymer. The determination of FTD is, however, very complicated. A unique method developed by us — chromatography of macromolecules at "critical conditions" close to the coil-adsorbed coil transition point — solves this problem for telechelic polymers. The basis of the method derived from the regularities of the adsorption of macromolecules is discussed and examples of separation of a number of hydroxyl-containing telechelic polymers with respect to the functionality types are given. Further development of the method for the analysis of other types of macromolecular heterogeneity is briefly discussed (the separation of cyclic and linear macromolecules is taken as an example).

1 Introduction 131

2 Functionality Type Distribution (FTD) — A Fundamental Characteristic of Telechelic Polymers 132
   2.1 Average Functionality, Functionality Type Distribution 134
   2.2 Classification of Oligomers According to their Functionality Types . . 135
   2.3 Methods of Determining the Average Functionalities 135
   2.4 Chromatographic Methods for FTD Determination 136

3 Theoretical Basis of the Liquid Chromatography of Macromolecules in the Critical Region 141
   3.1 Model of a Separating System 141
   3.2 Behaviour of a Macromolecule in a Limited Volume. Exclusion and Adsorption Modes. The Critical Region 143
   3.3 Chromatography of Macromolecules with Terminal Functional Groups 148
   3.4 Interrelation Between the Separation Modes in Liquid Chromatography and the Chemical Nature of the Macromolecule, the Stationary Phase and the Solvent 150
   3.5 Universal Dependences in the Critical Region 155

4 Separation of Telechelic Polymers in the Critical Region According to their Functionality Types 157
   4.1 Examples of Separating Hydroxyl-Terminated Polyesters and Polybutadienes 157
   4.2 Some Methodological Problems Concerning the Chromatography of Macromolecules in the Critical Region 166

5 Method for Analyzing Other Types of Macromolecular Heterogeneity 171

6 Conclusions 173

7 References 173

1 Introduction

The problem of molecular weight distribution (MWD) and functionality type distribution (FTD) belongs by definition to an extensive problem of *molecular heterogeneity* of polymers. In the synthesis of a polymer with the requested properties, e.g. a telechelic polymer, one is always faced with different types of polydispersity: the macromolecules can be of different length, they can have a different number of functional groups, i.e. be mono-, bifunctional, etc., they can be branched (star-, comb- or tree-like); and, finally, they can be cyclic.

These *heterogeneities*, which can be called "elementary", can be superimposed one on the other, i.e. bifunctional molecules can be linear or branched, linear molecules can be mono- and bifunctional, etc. In order to characterize in an ideal way a telechelic polymer with respect to its subsequent transformation, it is necessary to know a set of functions $\{f_i(M)\}$, the molecular weight distributions within each heterogeneity type. Clearly, it is very difficult in a general case to solve this characterization problem.

Using the traditional spectral methods, such as infrared or NMR spectroscopy, one can determine a type of heterogeneity present in the sample, such as the type of functional group or the existence of branch points. While the determination is complicated for long molecules because of a low concentration of functional groups, it does not yield information on how these heterogeneities are distributed among the macromolecules, i.e. how many macromolecules have, say, one, two, ... etc. functional groups.

On the other hand, when using methods sensitive to the size of macromolecules, e.g. exclusion chromatography, one runs into no lesser difficulties in interpreting the experimental results. Indeed, the dimensional distribution of macromolecules, obtained by exclusion chromatography, can be unambiguously correlated with MWD only within one heterogeneity type. However, for a sample consisting of e.g. a mixture of mono- and bifunctional molecules, the distribution obtained represents a sum of dimensional distributions of molecules having a different functionality and, therefore, cannot be attributed to a specific functionality type without additional assumptions. Combined detection does not solve the problem in a general case either.

Dimensional heterogeneity is thus seen to be closely connected with all other types of molecular heterogeneity. Without knowing the dimensional distribution, one cannot say in what way the heterogeneities are distributed among the molecules, and without knowing this distribution, one cannot unambiguously interpret the data of exclusion chromatography.

To resolve these problems, it is necessary to have a method sensitive both to the size of macromolecules and to the heterogeneities.

In the case of adsorption chromatography in contrast to exclusion chromatography, the macromolecules interact with the stationary phase, which makes the method sensitive to both the size of the macromolecule and its local structure, since the interaction of monomer units with the surface is restricted to the distance of several Angstrøms. It is one of the ways of solving the above-mentioned problems. This review presents the principal ideas underlying the method of chromatographic separation of macromolecules at critical conditions close to the coil-adsorbed coil transition point, where the heterogeneities manifest themselves most distinctly.

All the presented material is illustrated by the separation of linear functional and non-functional molecules. This is the simplest type of heterogeneity, but the solution of this problem is of great practical significance. The number of functional groups in a telechelic oligomer exerts a strong influence on the physico-chemical properties of the polymeric products. The knowledge of FTD is, therefore, important both for the synthesis — to optimize the production of the oligomer with the target functionality — and for its subsequent transformation into an elastomer with preassigned characteristics.

In Section 2, the concepts of functionality in the case of macromolecules and of the functionality type distribution are described in greater detail, and some examples are given of early FTD determinations using different chromatographic techniques.

Section 3 describes theoretical fundamentals of the new method developed by us — liquid chromatography of macromolecules at critical conditions (CCC). The theoretical ideas are illustrated by experimental data.

Some examples of macromolecule separation in critical conditions are examined in Section 4; a number of methodological questions are also discussed there.

Finally, Section 5 briefly describes the development of the method for the analysis of other types of heterogeneity.

2 Functionality Type Distribution (FTD) — A Fundamental Characteristic of Telechelic Polymers

Oligomers with reactive functional groups have been extensively applied to obtain a great variety of polymeric materials.

In contrast to polymers of a high degree of polymerization, the behaviour of oligomers is largely dependent on the nature and number of terminal groups. To characterize an oligomer, it is necessary to have information, not only about its composition and molecular weight but also its functionality.

The concept of functionality, which is clear and selfevident when applied to low-molecular-weight organic compounds, requires a special explanation when applied to polymers. When oligomers are synthesized, in addition to macromolecules of required functionality, functional-defective molecules are formed. They have, as a rule, their own MWD and a smaller or larger number of functional groups than required, and, sometimes, no such groups at all (Fig. 1). The concept of functionality and molecular weight has, in the case of high-molecular-weight compounds, a statistical meaning.

A vast majority of oligomers with functional groups, used as initial products in various polymerization processes, have not only a molecular weight but also a functionality type distribution. The term FTD — the relative content of macromolecules of different functionality — was for the first time introduced by Entelis [1]. It is imperative that an efficient synthesis of oligomers with the target functionality should be accompanied by a control of their FTD.

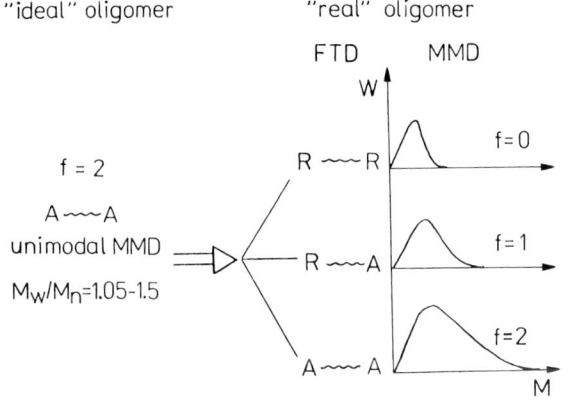

Fig. 1. Molecular weight distribution (MWD), and functionality type distribution (FTD) of a telechelic oligomer

One can say that theoretical and experimental studies have led to a conclusion on the dominant role of FTD in the properties of the end product.

Deviation of the average functionality of initial oligomers from the pre-assigned one can result either in a decrease or increase in the cross-linking density, caused by change in the number of side chains or linear molecules. Rather marked changes are observed in the kinetic parameters of the process (the content of sol, gelation time, critical conversion of the functional groups at the gel point, etc.), as well as in a number of properties of the final products.

Table 1 presents as an example the experimental dependences of kinetic parameters of the reaction, structural parameters of the network and some physico-mechanical characteristics on the molar fraction of monofunctional molecules in the reactant mixture, which is specific of FTD in the given case. As can be seen, the presence of monofunctional molecules, even in minor amounts (only 3–10%), exerts a strong influence on the kinetic and physico-mechanical parameters.

Table 1. Dependence of the kinetic parameters of the reaction, the structural parameters of the network and the physico-mechanical properties of polyurethane elastomers on the molar fraction of OH groups in the monofunctional reagent [2]

$[OH]_0$ g · eq/l	ϱ_3	$\varrho_1 \times 10^2$	γ_c (h)	η_c	$v/V \times 10^4$ mole/cm^3	σ kg/cm^2	ε %	E kg/cm^2
2.01	0.28	2.8	72	0.91	1.94	12.0	226	10.0
1.95	0.27	4.4	85	0.92	1.88	11.2	233	7.9
1.91	0.27	5.7	104	0.93	1.34	9.3	264	7.0
1.79	0.26	8.6	138	0.94	0.99	6.9	420	3.6
1.73	0.26	10.8	190	0.96	0.86	5.9	532	1.8

The reaction system consists of a bifunctional tetrahydrofuranpropylene oxide (THF-PO) copolymer, $M_n = 1700$, $f_n = 1.9$, and a monofunctional THF-PO copolymer, $M_n = 1400$, $f_n = 1$, trimethylolpropane and 2,4-toluylene diisocyanate; $[NCO]_0 = [OH]_0$.
ϱ_1 and ϱ_3 are the molar fractions of OH groups in mono- and trifunctional reagents, respectively, γ_c and η_c are the time and the degree of conversion at the gel point, respectively, v/V is the cross-linking density, σ, ε, and E are the strength, elongation at break and the tensile modulus, respectively.

2.1 Average Functionality, Functionality Type Distribution

Molecular or structural functionality, f, of any chemical compound and, specifically, a telechelic polymer should be understood as the number of functional groups in one molecule. If some of the functional groups are inactive, the practical or realizable functionality, f_p, of this compound is lower than its molecular functionality: $f_p < f$.

Macromolecules can differ in the number of reactive functional groups as well as in their nature. Also of importance is the number of bonds that can be formed by one functional group. It is, therefore, expedient to subdivide reactive functional groups (RFG) into single-act groups producing one bond, e.g. —OH, —COOH, —SH, —NH$_2$, —SO$_2$OH, —COCl, —SO$_2$Cl, etc., and two- and polyact groups, reacting with the formation of two and more bonds, e.g.

$$>C=C<, \quad -C\equiv C-, \quad >\!\!\underset{O}{\underset{\diagdown\diagup}{C-C}}\!\!<, \quad -\underset{O}{\overset{\|}{C}}-O-\underset{O}{\overset{\|}{C}}-, \quad -N=C=O.$$

For a linear polymer to be synthesized by the polycondensation or polymerization mechanism, each molecule of the oligomer must have f = 2 for single-act or f = 1 for two-act RFG. To obtain cross-linked polymers the system must contain components of f > 3 for single-act RFG (three-dimensional polycondensation) or with f > 2 for two-act groups.

In the case of oligomers, the concept of number-average functionality, f_n, is usually applied, which is the ratio between the total number of functional groups and the total number of molecules in the system, or, in other words, the average number of RFG per initial oligomer molecule.

The value of f_n is experimentally determined from

$$f_n = M_n/M_{eq}, \tag{2.1}$$

where M_n is the number-average molecular weight and M_{eq} is the equivalent molecular weight (average weight of the molecule per one functional group).

The f_n value provides information on the average functionality of oligomers, but does not characterize at all their functional polydispersity. For instance, $f_n = 2$, in some cases an ideal indicator for a telechelic oligomer, can be simulated by the presence of an equal number of mono- and trifunctional macromolecules.

To exclude this ambiguity and characterize the width of functionality type distribution it is possible to use, by analogy with average molecular weights, the values of number-average and weight-average, f_w, functionality:

$$f_n = \frac{\sum n_i f_i}{\sum n_i}; \quad M_n = \frac{\sum n_i M_i}{\sum n_i}$$

$$f_w = \frac{\sum n_i f_i^2}{\sum n_i f_i}; \quad M_w = \frac{\sum n_i M_i^2}{\sum n_i M_i} \tag{2.2}$$

where n_i is the number of molecules of functionality f_i.

For oligomers containing only one type of molecules, $f_w/f_n = 1$; in the case of a distribution of molecules of different functionality, $f_w/f_n > 1$.

It should be noted that for experimental determination of f_w (unlike that of f_n) there exists no direct method. Moreover, characterization of a telechelic polymer with the help of f_n and f_w is not sufficient and it is necessary to determine the entire FTD function.

2.2 Classification of Oligomers According to their Functionality Types

The simultaneous use of FTD and MWD to characterize oligomers enables subdivision of all known oligomers into three basic types.

Type 1 — oligomers with a strictly defined target functionality. In an ideal case, oligomers of this kind must have $f_w/f_n = 1$. They are usually synthesized by different methods: special techniques of initiation, telomerization, etc. In practice, however, oligomers of this type almost invariably are polydisperse in functionality, i.e. $f_w/f_n > 1$. The reasons and sources of functional defectiveness in different methods of synthesizing oligomers are briefly discussed in the review [1].

Type 2 — polyfunctional linear or branched oligomers with a regular alternation of functional groups along the chain. Oligomers of type 2 are characterized by a linear relationship between their molecular weight, M_i, and their functionality, f_i, as a result of which at fixed M_n the M_w/M_n value is egual to f_w/f_n. The appearance of FTD in such systems can be caused both by molecular-weight polydispersity and by the functional defectiveness resulting from the formation of cyclic structures or gels, as well as a consequence of deactivation of functional groups owing to the side reactions taking place during synthesis. Deviation of the experimentally measured f_n vs. M_n and f_w/f_n vs. M_w/M_n dependences from those calculated on the basis of the anticipated structural formula of the oligomer can serve as a quantitative characteristic of functional defectiveness.

Type 3 — polyfunctional linear or branched oligomers with irregular alternation of functional groups in the chain. These oligomers can have most diverse values of functional and molecular-weight polydispersity. One cannot speak of functional defectiveness of oligmers of this type without knowing their exact composition and compositional distribution. Investigation of FTD of such oligomers makes it possible to obtain information not only on molecular-weight and functional polydispersity but also on the compositional distribution.

Each of the above oligomer types obviously calls for a specific approach to choosing the methods of FTD determination.

2.3 Methods of Determining the Average Functionalities

Two approaches can be used for the experimental determination of average functionalities.

Indirect methods are mostly applied to determine f_w of oligomers. Successful application of these methods presupposes knowledge of the mechanism of transformation for the oligomers or the physico-mechanical properties of the resulting polymers obtained on their basis.

The most widely used indirect method at present is that of determining $(f_w)_A$ of oligomers with RFG from the gel point, depending on the weight-average functionality of the cross-linking reagent $(f_w)_B$, when oligomers are converted into cross-linked polymers via polycondensation or polyaddition reactions [3-8].

A modified Equation of Stockmayer [9] is used. With some changes in notation it has the following form:

$$(f_w)_A = 1 + r/\{P_A^2[(f_w)_B - 1]\} = 1 + 1/\{rP_B^2[(f_w)_B - 1]\} \qquad (2.3)$$

where

$$(f_w)_A = \frac{\Sigma (f_i)_A^2 A_i}{\Sigma (f_i)_A A_i}, \qquad (f_w)_B = \frac{\Sigma (f_i)_B^2 B_i}{\Sigma (f_i)_B B_i}$$

A_i and B_i are the numbers of molecules with functional groups A and B, $(f_i)_A$ and $(f_i)_B$ are their molecular functionalities, $r = A_0/B_0$.

To determine $(f_w)_A$ from Eq. (2.3), one must know the weight-average functionality $(f_w)_B$ of the cross-linking reagent, the A_0/B_0 ratio and the conversion at the gel point (P_A or P_B). The gel point is usually determined from viscosity changes of the reaction system.

There also have been indications in the literature that it is possible to determine f_n of the initial prepolymer using the dependence of cross-linking density [7,10] and the Mooney-Rivlin constant [10] for resulting polymers on the functionality, f_n, of initial reagents. Indirect methods require an independent calibration whose unambiguousness must be proved in each particular case. They are mainly applied to oligomers of type 1 and, exceptionally, of types 2 and 3 when the functionality is not high, $f \sim 3 - 5$ [7].

In the literature only a few cases have been reported of experimental f_w determination for oligomers with RFG by their gel point, mainly referring to estimating the accuracy of the method on model systems [6-8], finding f_w for oligobutadienes with carboxyl and hydroxyl groups [3-5], and poly(oxypropylene) triol and diol with hydroxyl groups [5a].

Direct determination of f_n from the ratio of the number-average molecular weight, M_n, to the equivalent molecular weight, M_{eq}, is applicable, with certain restrictions, to all the above three types [Eq. (2.1)].

2.4 Chromatographic Methods for FTD Determination

The use of liquid chromatography to investigate the functionality of oligomers was independently proposed by us [11] and by Muenker and Hudson [12]. Column fractionation of poly(diethylene glycol adipate) (PDEGA) filled with activated silica gel showed that, apart from the separation of oligomer macromolecules according to their molecular weight also separation according to the content of terminal hydroxyl groups takes place [11]. This allows separation of oligomers of a similar molecular weight according to their functionality. Muenker and Hudson [12] used mixtures of solvents with increasing fractions of components having a high elution ability to separate hydrogenated oligobutadienes with hydroxyl and carboxyl terminal groups according to their functionality types. A practically complete separation of non-, mono- and bifunctional molecules could be performed.

Later, the method of liquid adsorption chromatography, both in the isocratic and the gradient separation modes, was applied to investigate the FTD of oligomers, mainly of the first and the third types, with —OH, —COOH, —C=N and —SH functional groups.

Table 2 presents some experimental results obtained by separating oligomers according to their functionality or investigating the influence of functional groups on their chromatographic behaviour. The data were obtained by column liquid or thin-layer chromatography. An overwhelming majority of oligomers used in industry were shown to contain "defective" molecules along with macromolecules of target functionality.

Bifunctional oligobutadienes were found to contain, besides bifunctional molecules, non- and monofunctional molecules [16, 17, 19], and from a series of polyesters and thiokols fractions of non-functional molecules of cyclic structure were isolated [39, 45]; poly(oxypropylene) with target functionality $f \geq 3$ contained fractions of mono- and bifunctional macromolecules [26, 27, 29, 34], etc.

The data in Table 2 can be summarized as follows (Fig. 2): For some separation systems, the functional groups do not manifest themselves in any way, separation proceeds similarly as in conventional exclusion chromatography, the calibration curves for non-functional and functional molecules coincide, and the MWD obtained is the sum of MWD of molecules with different functionality. Such MWD is only of relative

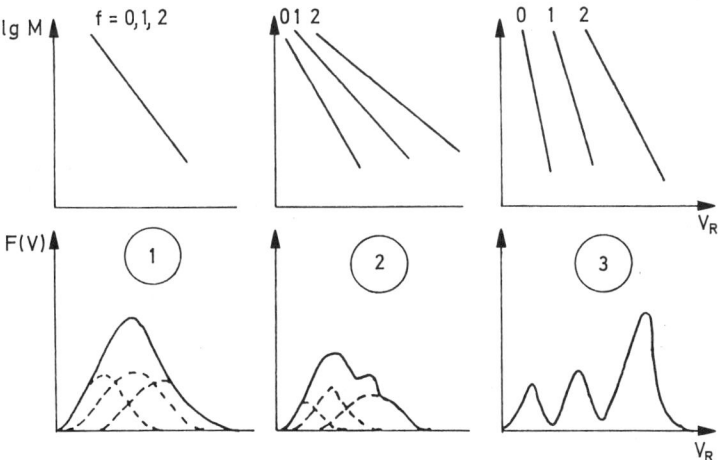

Fig. 2. The most typical experimental situations that can be encountered in chromatography investigating the FTD and MWD of oligomers.
1: functional group does not manifest itself chromatographically, 2: a partially resolved chromatogram, 3: a fully resolved chromatogram when a simultaneous FTD and MWD determination is possible

Table 2. The use of liquid absorption chromatography to investigate the FTD of functional oligomers

	Oligomer	Functional group	M_n	f_n	Stationary phase	Eluent (separation conditions)	Method	Ref.
1.	Oligobutadiene (hydrogenated)	OH	2000	2	silica gel	carbon tetrachloride — chloroform — ethanol (g)	LC	10, 12)
2.	Oligobutadiene	OH	2000–6000	1.9–2.3	silica gel		LC	12–14)
			2000–4000	1.95–2.0	silica gel	chloroform — ethanol (g)	LC	15)
3.	Oligobutadiene (hydrogenated)	COOH	3000	2.0	silica gel	carbon tetrachloride — chloroform — ethanol (g)	LC	12)
4.	Oligobutadiene	COOH			silica gel	carbon tetrachloride — chloroform — ethanol (g)	LC	13, 14, 16)
			1000–6000	0.5–2.05	silica gel	chloroform — acetone — ethanol (g)	LC	15, 17)
5.	Oligoisobutylene	COOH	2500	2	silica gel	carbon tetrachloride — chloroform — ethanol (g)	LC	12)
			2200	1.6	silica gel	chloroform — acetone (g)	LC	17)
6.	Oligoisoprene	OH	900	1.8	silica gel	carbon tetrachloride — chloroform — ethanol (g)	LC	17)
			3000–10000	0–2	modified silica NH_2	hexane-dichloroethane, 80/20 (i), hexane-isopropyl alcohol, 97/3 (i)	LC	17a)

	Oligomer	Functional group	M_n	f_n	Stationary phase	Eluent (separation conditions)	Method	Ref.
7.	Block copolymer of butadiene-1,3 and isoprene	OH	3100	1.6	silica gel	carbon tetrachloride — chloroform — ethanol (g)	LC	14, 18)
			1200–2700	1.8–1.9	silica gel	carbon tetrachloride — chloroform, 40/60 (i)	LC	19, 20)
			2800–3100	1.3–2.0	silica gel	carbon tetrachloride — chloroform — acetone — ethanol (g)	LC	17)
8.	Copolymer of tetrahydrofuran with propylene oxide	OH	1000–1300	0.8–2.0	silica gel	methylethyl ketone (i)	LC	21, 22)
9.	Poly(tetrahydrofurane)		500	0–2	silica gel	methylethyl ketone (i)	LC	21, 23, 24)
10.	Homopolymer of propylene oxide	OH	4000–5000	2–3	silica gel	hexane — methylethyl ketone — ethanol, 6:2:1 (i)	TLC	25)
11.	Polyoxypropylene polyols	OH	400–3000	1–5	silica gel	ethyl acetate saturated with water (i)	TLC	26–28)
			250–6000	1–5	silica gel	ethyl acetate saturated with water — methylethyl ketone (i)	LC	29–35)
			400–3000	1–5	silica gel	ethyl acetate saturated with water — methylethyl ketone (g)	LC	27, 28, 34)
12.	Poly(tetramethylene glycol)	OH	500–3000	1–2	silica gel	methylethyl ketone (i)	LC	36)

Table 2 (Continued)

13.	Poly(diethylene glycol adipate)	OH	400–2000	0–1.8	silica gel	methylethyl ketone (i)	LC 11, 37)
			900–2400	1.70–1.95	silica gel	hexane — methylethyl ketone — ethanol (g)	LC 38–40)
		COOH	1500	1.8	silica gel	hexane — methylethyl ketone — ethanol (g)	LC 39)
14.	Poly(butylene glycol adipate)	OH	1000–4000	0–2	silica gel	benzene — ethanol, 75/25 (i)	TLC 41)
			2000	2	silica gel	toluene — ethanol (g) toluene — ethanol, 93/7 (i)	LC TLC 42)
15.	Poly(butylene ethylene glycol adipate)	OH	2000	2	silica gel	toluene — ethanol (g) toluene — ethanol, 93/7 (i)	LC TLC 42)
16.	Copolymer of butadiene-1,3 with acrylonitrile	COOH CN	1500	2	silica gel	carbon tetrachloride — chloroform — ethanol (g)	LC 14, 16)
17.	Copolymer of butadiene-1,3 with metacrylic acid	COOH	2300	1.9–2.0	silica gel	carbon tetrachloride — chloroform — ethanol (g)	LC 14, 16)
18.	Copolymer of butadiene	COOH	3000	1.9	silica gel	carbon tetrachloride — chloroform — ethanol (g)	LC 13)
			2900	2.0–2.5	silica gel	chloroform — ethanol (g)	LC 43)
19.	Epoxide resins		350–1000	1.8–2	styrogel, butylmetacrylate gel	tetrahydrofuran (i)	GPC 44)

value, since the behaviour of an oligomer as a reagent is determined not by the overall MWD but by MWD in the zones of different functionality. Another case, most often encountered, is that when a functional group manifests itself only slightly, the calibration curves for different functionalities do not coincide, but, because of MWD, the zones of different functionalities overlap. In this case, only a qualitative characterization of polydispersity in the functionality of a sample is possible. Finally, in the cases when the interaction of a functional group with the adsorbent is much stronger than that of the backbone units (e.g. in polybutadienes with hydroxyl and carboxyl groups) one can achieve complete separation of the zones of different functionality.

In recent years, owing to the rapid development of experimental techniques, great progress has been made in liquid chromatography in separating complex mixtures. It has become necessary to devise a method utilizing all the possibilities of modern high-performance columns and chromatographs capable of feeding the solvent into a column at a high pressure, with precise control of the flow rate and the composition.

From the studies listed in Table 2 it is clear that all the separation methods are based on a change in the functional group — adsorbent interaction affecting by a change in the solvent composition which is accompanied by a change in the interaction of backbone units. It is, therefore, necessary to understand the regularities of the chromatography of linear macromolecules, not necessarily functional, capable of interacting (as distinct from exclusion chromatography) with the stationary phase. The length and flexibility of macromolecules make their behaviour in adsorption chromatographic separation qualitatively different from that of low-molecular substances. Further on, the emphasis will be placed on the use of high-performance chromatography for the analysis of the FTD of macromolecules.

3 Theoretical Basis of the Liquid Chromatography of Macromolecules in the Critical Region

3.1 Model of a Separating System

Any chromatographic process is associated with the distribution of the analyzed substance between the mobile and the stationary phase. In liquid chromatography the solvent, with the volume V_0 in the interparticle space, moving along the column at a certain speed, is the mobile phase, and a porous adsorbent, with an overall pore volume of V_p, is the stationary phase. The distribution coefficient, K_d, equal to the ratio between the concentration of the substance in the stationary and the mobile phase, determines the retention volume, V_R, of a given substance in the column according to the basic chromatographic equation

$$V_R = V_0 + V_p K_d \tag{3.1}$$

K_d is related to the free energy change at the point where the macromolecule passes from the mobile into the stationary phase (Fig. 3), where F_o and F_p are the free energies in the solution and in the pore:

$$K_d = \exp(-\Delta F)\ ^1$$

Assuming $F_o = 0$, we get $\Delta F = F_p$.

A vast number of theoretical studies have been devoted to the behaviour of macromolecules in restricted volumes (in our case a macromolecule in the stationary phase pore). Examined are both the "ideal" and the "real" chains in the pores of different geometry with a different character of the interaction (repulsive or attractive) of units with the surface [46-50]. An analysis of these studies and their applicability to solving the specific problems goes far beyond the scope of the present review [51, 52]. We only note that in a great majority of these studies either repulsing or attracting walls are considered, whereas we are interested in the coil-adsorbed coil transition. Moreover, of interest for chromatography are mainly those studies where an isolated macromolecule is dealt with. Therefore, for simplicity the conclusions will be formulated for a Gaussian model macromolecular chain examined in [53-55]. Sometimes, a particular assertion will be illustrated by presenting the macromolecule in the form of random walk on a lattice of a mesh size a coinciding with the size of the segment. Strictly speaking, this approach is only applicable to long coil-like molecules; however, as it will be experimentally shown below, the notions developed for long molecules do not lead to qualitative errors if one analyzes the chromatographic behaviour of short flexible oligomer chains.

The main object of the theory is to find from the analysis of the simplest model which of free-energy parameters are and which are not significant. It is also

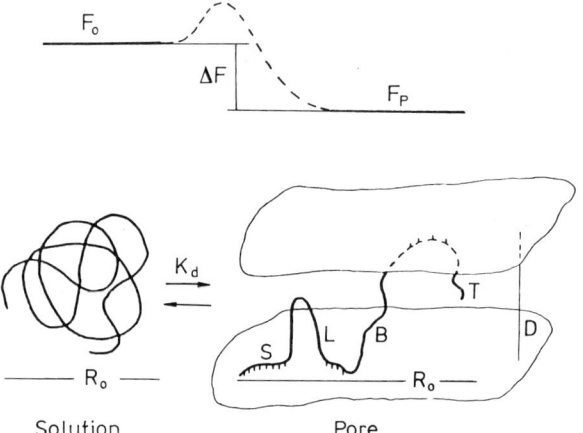

Fig. 3. Schematic picture of a macromolecule in the adsorbent pore. S sequence, L loop, B bridge, T tail of macromolecule

---

1 All energy values will be expressed in kT units

necessary to establish the relationship of significant parameters and the chemical nature of the macromolecule and separating system (composition of eluent, type of stationary phase, etc.). This relationship is most conveniently established within the framework of some kind of correlative approach.

3.2 Behaviour of a Macromolecule in a Limited Volume. Exclusion and Adsorption Modes. The Critical Region

In its general formulation, the problem of a macromolecule in a limited volume is the problem of a macromolecule in an external field [55]. The solution of this problem is well known: the partition function of the chain starting at point r and ending at point r' has the form of bilinear expansion [53, 55],

$$G_N(r, r') = \Sigma \Lambda_i^N \psi_i(r) \psi_i^+(r') \tag{3.2}$$

Λ_i and ψ_i are the eigenvalues and the corresponding eigenfunctions of the integral operator $\hat{g}$,

$$\hat{g}\psi_i = \int g(r - r') \psi_i(r') \, dr' = e^{\varphi(r)} \Lambda_i \psi_i(r) \tag{3.3}$$

The function $g(r - r')$, the so-called "linear memory" function [55] describes the bonding of units into a chain, $\varphi(r)$ is the external field into which the macromolecule is placed. In a limited volume, the spectrum of eigenvalues in Eq. (3.3) is discrete; therefore, for sufficiently high N

$$G_N(r, r') \approx \Lambda^N \psi(r) \psi(r') , \tag{3.4}$$

where Λ is the highest eigenvalue and ψ is the corresponding eigenfunction. As we see, the ends of a macromolecule become in this case statistically independent. For this reason, it is necessary that the size of the macromolecule coil, R_0, should be larger than the characteristic size of pores D of the stationary phase.

Let us consider a slit-like pore of width D along whose walls the $\varphi(x)$ potential is localized (Fig. 4). We shall regard the interaction of monomers with the walls as a short-range interaction and the characteristic radius of interaction as being of the order of the segment size a. The exact assignment of the form of the potential is immaterial for our purposes, since it describes the effective interaction of units with the pore walls, renormalized by the solvent molecules. Conditions are to be as follows:

$$\varphi(x) = \begin{cases} \infty, & |x| > D/2 \\ -\theta, & -D/2 < x < -D/2 + a; \quad D/2 - a < x < D/2 \\ 0, & |x| < D/2 - a \end{cases} \tag{3.5}$$

In most cases of practical importance (the size of sorbent pores commonly used in practice lies within the range of 60–4000 Å), it seems possible to disregard the size of interaction a in comparison with D and reduce the action of the potential to

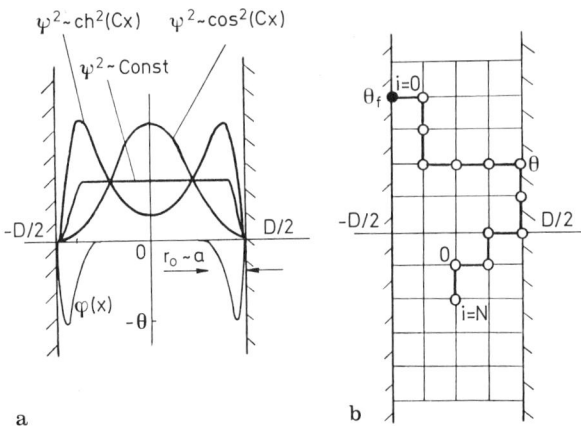

Fig. 4a and b. Distribution of the segment density at different values of the energy θ (**a**) and schematic picture of lattice-like chain of length N in a slit-like pore of width D (**b**). θ_c is the critical energy characteristic of the case when the entropy losses of the macromolecule in the pore are compensated by the energy of interaction with the wall. φ(x) is the attractive potential of a depth θ and with a characteristic radius of interaction r_0 of the order of the segment size a

different, in the case of attraction or repulsion, boundary conditions. In addition, at $D \gg a$ the integral operator is reduced to a differential one [50, 55]

$$\hat{g}\psi \to \left(1 + \frac{a^2}{6}\frac{d^2}{dx^2}\right)\psi.$$

This operation very much simplifies the finding of eigenvalues and the corresponding eigenfunctions. To find them it is necessary to solve a differential equation in the following form:

$$\frac{a^2}{6}\frac{d^2}{dx^2}\psi = (\Lambda - 1)\psi \tag{3.6}$$

with different boundary conditions in the case of attraction or repulsion.

Integration of Eq. (3.4) with respect to the volume of the pore yields the statistical sum for the macromolecule in the pore

$$Z_N = \Lambda^N \int_{V_p} \psi(r)\,\psi(r')\,dr\,dr' \tag{3.7}$$

From Eq. (3.7) it follows that

$$F_p = -N \ln \Lambda - 2\ln\left\{\int_{V_p} \psi(r)\,dr\right\} \tag{3.8}$$

If $\Lambda \neq 1$ and for long chains, the logarithmic term in Eq. (3.8) can be disregarded. However, if for some reason $\Lambda = 1$ or $\Lambda \to 1$, one can no longer disregard the logarithmic term, and its contribution to the free energy F_p is not small. The whole difference between non-functional molecules and molecules with terminal functional groups resides in the logarithmis term of Eq. (3.8) and manifests itself only under the conditions at which this term cannot be disregarded. The $\psi(r)$ function is proportional to the probability density of finding the end of the chain at point r. The energy of the terminal functional groups' interaction with the pore surface, θ_f, being different from the interaction energy of the ends of a non-functional chain, θ, is assumed equal to the interaction energy of backbone units, the probability of finding the ends of a functional molecule close to the pore wall differs from that for a non-functional molecule.

As it will be shown below, $\Lambda \approx 1$ corresponds to the region lying close to the critical conditions, the region where the macromolecule passes from the solution into the stationary phase pore.

Let us first examine the behaviour of macromolecules in the case of repulsing walls (we shall call this case the exclusion mode) and in the case of attracting walls (the adsorption mode).

a. The case of repulsion

The boundary conditions for repulsing walls [50] are as follows:

$$\psi|_{x=\pm D/2} = 0$$

Therefore,

$$\Lambda = 1 - \frac{\pi^2 a^2}{6D^2}, \quad \psi = \text{const} \cdot \cos\left(\frac{\pi x}{D}\right)$$

and the free energy

$$F_p \approx -N \ln\left(1 - \frac{\pi^2 a^2}{6D^2}\right) \approx \frac{\pi^2}{6} \frac{Na^2}{D^2}$$

This is a well-known result obtained by Casassa [46]. In the exclusion mode, the longer the macromolecule the greater is the energy that has to be spent to place it in the pore. The distribution coefficient K_d exponentially decreases to zero with increasing size of the macromolecule. The larger the molecule, the smaller is V_R in accordance with Eq. (3.1); in the exclusion mode $\partial V_R/\partial N < 0$.

b. The case of attraction

In this case, the boundary conditions for $\psi(r)$ can be written as follows [56]:

$$\frac{1}{\psi} \frac{d\psi}{dx}\bigg|_{x \approx \pm D/2} = \pm k(\theta)$$

Here $1/k > 0$ and by the order of magnitude it coincides with the adsorption layer thickness; it is determined by the exact solution of Eq. (3.3) close to the pore walls. Let us consider the case with small k; only such values are meaningful for chromatography, since otherwise the adsorption of macromolecules becomes irreversible. For small k

$$\Lambda \approx 1 + \frac{ka^2}{3D}, \quad \psi = \text{const} \cdot \cosh\left(\sqrt{\frac{2k}{D}}\,x\right),$$

and the free energy of a macromolecule in the pore

$$F_p \approx -N \ln\left(1 + \frac{ka^2}{3D}\right) \approx -k\frac{Na^2}{3D}.$$

For attracting walls on the contrary, the larger the macromolecule the longer it stays in the stationary phase pores, K_d grows exponentially with increasing molecular weight, and $\partial V_R/\partial N > 0$.

Changing the depth of the potential θ, e.g. by changing the composition of the solvent or the temperature (since θ is in kT units), one can find a certain value of θ_c at which k = 0. We shall call such conditions critical. They correspond to the case when the entropy losses of the bond between two successive monomers close to the pore wall, at a distance of the order of the segment size a, are compensated by the interaction energy with the wall.

Let us consider, as an example, a polymer chain on the Flory-Huggins lattice. If the i-th monomer has got onto the pore wall, for the (i, i − 1)-th bond there are only 5 possible states instead of 6 in the unrestricted volume. For the entropy losses, ΔS = ln 6/5, to be compensated, we must set for the interaction energy of the i-th monomer [48]

$$\theta_c = \ln 6/5 \approx 0.2.$$

This value of critical energy for a cubic lattice will be used henceforth.

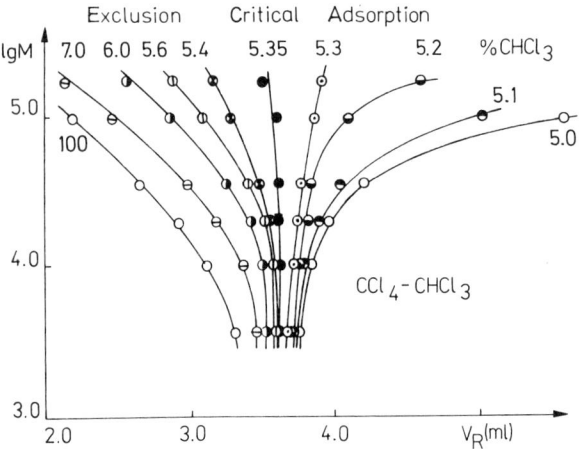

Fig. 5. Transition from the exclusion to the adsorption separation mode through critical conditions for polystyrene standards at a varying composition of the binary eluent (CCl$_4$–CHCl$_3$). (Column Si-300, flow rate u = 0.5 ml/min, volume of sample 10 μl, UV detector, λ = 275 nm, t = 27 °C)

In critical conditions (CC), the highest eigenvalue and the corresponding eigenfunction are equal to:

$$\Lambda = 1, \quad \psi = \text{Const}$$

The existence of critical conditions, separating the exclusion and the adsorption modes, is a fundamental fact in the chromatography of macromolecules. As far as we know, the transition from the exclusion to the adsorption mode within the framework of a single mechanism of the chromatography of macromolecules was investigated for the first time, using the methods of column and thin-layer chromatography, in Refs. [57] and [58]. The following figures illustrate this phenomenon.

Figure 5 shows the transition through CC for polystyrene standards. The results are presented in the form of calibration curves in the traditional lg M — V_R coordinates.

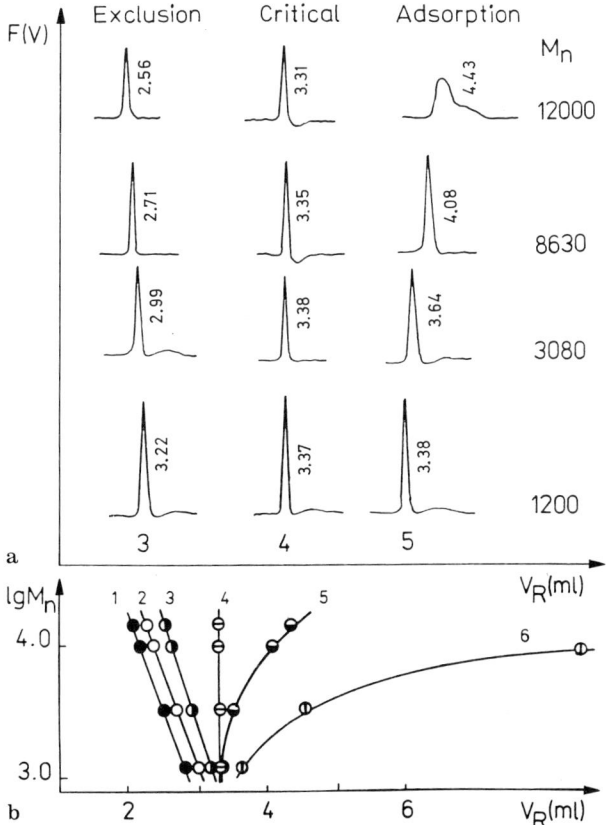

Fig. 6. Experimental chromatograms (a) and shape of calibration dependences (b) for nonfunctional oligobutadienes of different molecular weight in the exclusion, critical and adsorption separation modes.
(Column Si-100, eluent heptane-methylethyl ketone, u = 1 ml/min, volume of sample 10 μl, refractometer, t = 24 °C) % heptane 0 (1), 10 (2), 99 (3), 99,5 (4), 99,7 (5)

Oligobutadiene behaves in just the same way when the mobile phase composition changes (Fig. 6). There is no retention volume vs. molecular weight dependence in CC for either polystyrenes or butadienes.

3.3 Chromatography of Macromolecules with Terminal Functional Groups

Let us now consider the chromatography of functional macromolecules. Since the presence of functional groups results in the terminal units interacting with the surface differently than the chain units, statistical weights of such configurations of functional molecules that begin or end on the walls differ $\sigma = \exp\{\theta_f - \theta\}$ times from the respective configurations of non-functional molecules.

Let a functional group at the end of a molecule. The statistical sum for such a molecule in the pore can be written as

$$Z_N^{(1)} \approx \Lambda^N \int_{V_p} \psi_f(r) \, \psi(r') \, dr \, dr' \,.$$

Taking into account that

$$\int \psi_f(r) \, dr \approx \int \psi(r) \, dr \left\{ 1 + \frac{2(\sigma - 1) \int_{D/2-a}^{D/2} \psi(x) \, dx}{\int \psi(x) \, dx} \right\}$$

for monofunctional molecules we shall get

$$Z_N^{(1)} = Z_N^{(0)} \left\{ 1 + \frac{2(\sigma - 1) \int_{D/2-a}^{D/2} \psi(x) \, dx}{\int \psi(x) \, dx} \right\},$$

where $Z_N^{(0)}$ designates the statistical sum of Eq. (3.7) for a non-functional molecule of the same length N. Because of statistical independence of the chain ends (it is to be remembered that the case in question is that of $R_0 > D$), we get for a bifunctional molecule

$$Z_N^{(2)} = Z_N^{(0)} \left\{ 1 + \frac{2(\sigma - 1) \int_{D/2-a}^{D/2} \psi(x) \, dx}{\int \psi(x) \, dx} \right\}^2 \tag{3.10}$$

Expressions (3.9–10) acquire an especially simple form in CC when $\Lambda = 1$. Then, from the condition of normalization $\int_{V_p} |\psi|^2 \, dr = 1$ we can find that

$$\psi = \frac{1}{\sqrt{V_p}} \,. \tag{3.11}$$

Therefore, for non-functional molecules

$$Z_N^{(0)} = V_p$$

and the distribution coefficient $K_d = 1$.

If $R_0 < D$, from the solution of Eq. (3.6) and the general expression for G_N [see Eq. (3.2)] it is easy to see that also this case K_d is equal to unity. Thus, in CC, the retention volumes of non-functional molecules, in which all units have the same energy of interaction with the pore walls, depend neither on the macromolecule length nor on the size of stationary phase pores

$$V_R^{(0)} = V_0 + V_p \tag{3.12}$$

For mono- and bifunctional molecules, from Eqs. (3.9–11) we get [59, 60]

$$K_d^{(0)} = 1$$

$$K_d^{(1)} = 1 + \frac{2a}{D}(e^{\theta_f - \theta_c} - 1).$$

$$K_d^{(2)} = \{K_d^{(1)}\}^2 = \left\{1 + \frac{2a}{D}(e^{\theta_f - \theta_c} - 1)\right\}^2 \tag{3.13}$$

It follows from Eq. (3.12) that under critical conditions both the interparticle space volume and the total volume of pores are equally accessible to the macromolecule.

Then, $K_d^{(1)}$ represents the distribution coefficient for a point-like (small) molecule having one functional group.

From this analysis, the following conclusions can be drawn with respect to the chromatography of functional molecules. Both in the exclusion ($\theta < \theta_c$) and the adsorption ($\theta > \theta_c$) modes, for sufficiently long molecules one can disregard the logarithmic term in Eq. (3.8). Therefore, the functionality will not manifest itself since it enters into the statical sum only through this term. At critical conditions on the contrary, the logarithmic term cannot be disregarded at any chain lengths, and, therefore, the functionality (the difference in the interaction with the surface for only one group out of thousands) must manifest itself even for very long molecules. With a decrease in the length of the chain, $N \to 1$, the distribution coefficients for mono- and bifunctional molecules, both in the exclusion and the adsorption modes, must tend to their values in the critical conditions, since these are in fact the distribution coefficients for small molecules. As a result, in lg N — K_d coordinates have the picture shown in Fig. 7.

A detailed study of the chromatographic behaviour of the lattic-like models of chains in slit-like pores has been performed in Refs. [59–61], for all chain length N and pore size D ratios and all energies of the interaction of units with pore walls, θ. In Ref. [62], a strict analytical theory has been elaborated for the separation according to the functionality of macromolecules interacting with the surface only by their terminal groups.

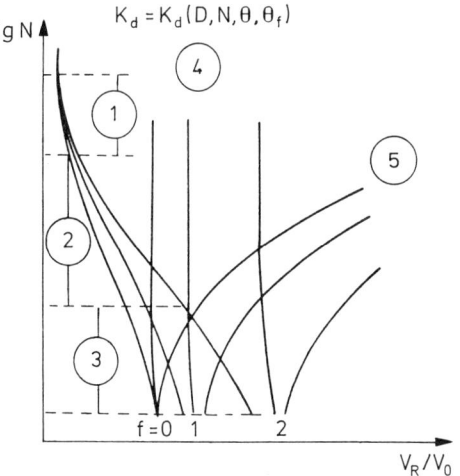

Fig. 7. Interrelation between molecular weight and retention volume for macromolecules of different functionality at chromatography in the exclusion (1–3), the critical (4), and the adsorption (5) separation modes [59]. In the general case, the distribution coefficient K_d is a function of the pore size D, the chain length N, the interaction energy with the pore wall of the backbone segments θ and the terminal segment θ_f containing the functional group (zones 1, 2 and 3 correspond to the cases shown in Fig. 2)

From Fig. 7, it is easily seen that critical conditions are the optimum conditions for separation according to the functionality. There is no molecular weight separation, and the chromatogram can be unambiguously interpreted as FTD.

Close to the critical conditions in the exclusion as well as in the adsorption mode, there also exists a region where both the functionality and the molecular weight separations are possible. However, the size of this region depends on the difference between the interaction energies of the terminal and the chain units $\theta_f - \theta$, the size of adsorbent pores D and on the specific form of MWD function. In the general case, in the exclusion and the adsorption modes the molecular weight separation is superimposed on the functionality separation, which makes an unambiguous interpretation of the chromatograms very difficult.

Before passing to the practical verification of the concepts presented above, we shall establish the interrelation between the separation modes and the chemical nature of the macromolecule, solvent (eluent) and stationary phase (adsorbent).

3.4 Interrelation Between the Separation Modes in Liquid Chromatography and the Chemical Nature of the Macromolecule, the Stationary Phase and the Solvent

As it was shown in Section 3.2, close to the critical conditions the distribution coefficient K_d is a function of chain length, pore size D and the energy of interaction of units with pore walls, θ. For a chosen molecule and adsorbent, $K_d = K_d(\theta)$, and, therefore, by changing θ one can successively achieve the transition from the adsorption to the exclusion mode and vice versa, finding in this way the critical conditions necessary for separation according to the functionality.

For a given macromolecule and adsorbent, there exist two ways of changing θ: by changing the solvent composition or the temperature. It is, however, inconvenient to use temperature variation, since the interaction of the units with the surface is renormalized by the presence of solvent molecules, and the $\theta(T)$ dependence can be of complex and non-monotonous nature. Such phenomena are well known in low-molecular-weight liquid chromatography [63]. It is more convenient to use the variation of the solvent composition.

In adsorption chromatography, the energy of the interaction of a certain substance X with the pore surface and the variation of retention volumes depending on the polarity of the mobile phase are commonly assessed using Snyder's correlative approach [64]. According to this approach, the energy of the interaction of substance X with adsorbent surface A from solvent S can be written as

$$\theta = E_{XA} - E_{SA} + E_{SS} - E_{XS} \tag{3.14}$$

Assuming further that the interaction of substance X with the solvent and of solvent molecules with one another is of the same nature, one can cancel the last two terms on the right-hand side of Eq. (3.14) and obtain the basic relationship

$$\theta \approx E_{XA} - E_{SA} \tag{3.15}$$

This relationship can be rewritten in terms of correlation parameters

$$\theta = \alpha(X_0 - A_x\varepsilon_0)$$

that have a simple physical meaning: X_0 is the energy of adsorption of substance X on a standard adsorbent, A_x is its molecular area, ε_0 is the solvent strength² on a standard adsorbent, α is the adsorbent activity, associated with the content of water in it. For a standard absolutely dry adsorbent $\alpha = 1$.

The X_0 and A_x values for a specific substance can be calculated from the additivity, proceeding from the X_i and A_i values for different groups included in this substance. Tables for such calculations are available in [64]. This problem has been comprehensively discussed, and accompanied with numerous examples in Snyder's monograph. We shall only note that the values of A_x are only exceptionally related to the concrete size of the molecule X and represent a certain independent variable.

In a first approximation, the energy of interaction of the segment with the surface, θ, can be assumed equal to the energy of the interaction of a low-molecular-weight analogue of the segment having the same chemical structure, and can be calculated from the tabular values of the energies of the groups constituting the segment and from the solvent strength ε_0 [60]. A unit of the polymer chain interacts with the surface amidst other units whose local concentration is not small, while the adsorption energy of the low-molecular-weight analogue, calculated from the adsorption energies of the groups, corresponds to the conditions when each molecule of the analogue interacts with the sample independently. Therefore, the bonding of units into a chain renormalizes the interaction energy with the surface for each unit.

---

2 ε_0 by its physical meaning coincides with the adsorption free energy of the solvent per 0.085 nm² of the surface

We have calculated the energy θ in this way for some polymers and separation conditions (Table 2) and, using the lattice-like model and a slit-like pore, we have found the distribution coefficients, $K_d^{(i)}$, for these macromolecules as a function of N, D, θ and $θ_f$ [65]. It turned out that for such a crude model not only the calculated $K_d^{(i)}$ values were close to the experimental ones, but also, which is especially important, that the chemical nature of the macromolecule, the functional groups and the separation conditions (the mobile phase composition) were correctly accounted for. Two examples of such calculations are given in Figs. 8 and 9.

Therefore, the interaction energy is reasonably expressed by Eq. (3.15) not only for low-molecular-weight substances but also for macromolecules, if the substance X is a macromolecule segment.

One of the principal advantages of Eq. (3.15) consists in the solvent power of a binary mixture a and b, $ε_{ab}$, being determined through their individual $ε_{0a}$ and $ε_{0b}$ values ($ε_{0a} < ε_{0b}$) and the molar fraction C_b of the polar component. General considerations lead to [64]

$$ε_{ab} = ε_{0a} + \frac{1}{αA_b} \lg[C_b\, 10^{αA_b(ε_{0b}-ε_{0a})} - C_b + 1] \qquad (3.16)$$

For a given polymer with fixed X_0 and A_X, Eqs. (3.15) and (3.16) describe a non-linear but monotonous variation of θ with the composition of the mobile phase C_b (Fig. 10). The procedure of finding the critical conditions then becomes very simple: it is necessary to find two solvents, in one of which (a) the adsorption and in the other one (b) the exclusion mode is operative and then, by changing their ratio, to find the point C_b^* where there is no retention volume dependence on the molecular weight of the polymer. The only requirement imposed on the choice of the solvents is that they both should be good solvents. In practice, however, it is

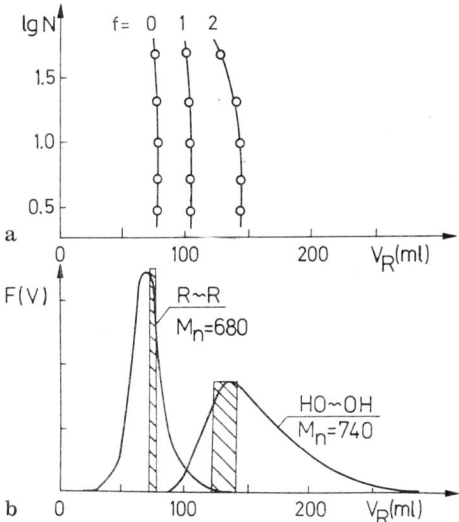

Fig. 8a and b. Calibration dependence (a) and chromatograms (b) of poly(diethylene glycol adipate) in methylethyl ketone [65]. (the cross-hatched regions indicate the calculated values of V_R for N = 5 — 15)

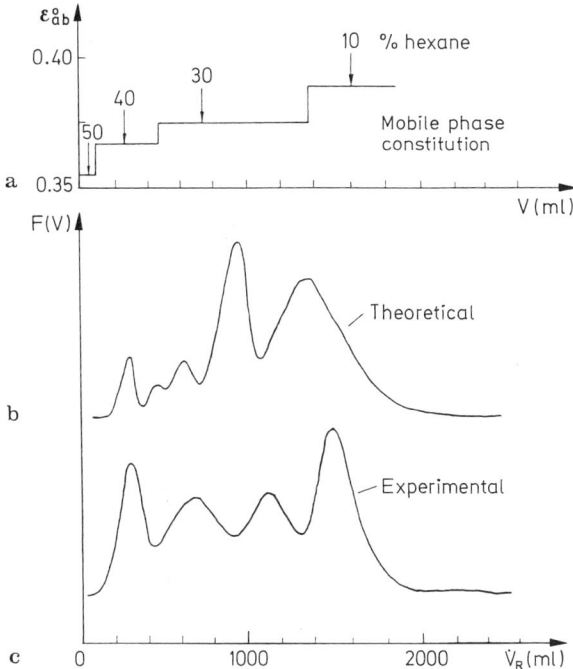

Fig 9a–c. Eluent programme $\varepsilon_{ab}(V)$ (eluent hexane-methylethyl ketone) (**a**); theoretically calculated (**b**), and experimental (**c**) chromatograms of poly(diethylene glycol adipate) of $M_n = 550, f = 1.63$ [65]

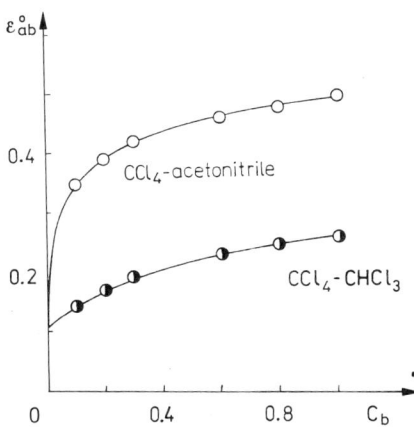

Fig. 10. Dependence of the chromatographic power of the binary solvent ε_{ab} on the molar fraction C_b of solvent b

sufficient for the mixture of solvents to be thermodynamically good close to C_b^*, where small changes in composition strongly affect K_d.

If one takes into account the binding of units in the chain, the values of X_0 and A_x, i.e. the energy of the interaction with the surface and the molecular area of the segment, respectively, have to be determined. These values easily can be determined by using the critical conditions [66]. In the close vicinity of the critical consitions and for the Flory-Huggins cubic lattice of a mesh size a, it is easy to find

$$\ln K_d \approx \frac{2aN}{D}(\theta - \theta_c) \qquad (3.17)$$

$$\theta_c = \ln 6/5$$

From Eqs. (3.15) and (3.17), it follows that, $\ln K_d = \ln K_d(\varepsilon - \varepsilon^*)$, and the slope of this plot at point $\varepsilon = \varepsilon^*$ is proportional to the molecular area A_X. X_0 and A_X are determined by Eqs. (3.18)

$$\left.\frac{\partial \ln K_d}{\partial \varepsilon}\right|_{\varepsilon^*} = \text{const } A_x \qquad (3.18)$$

$$\ln 6/5 = \alpha(X_0 - A_X\varepsilon^*)$$

We have performed such calculations for samples of non-functional polybutadienes [66] (Fig. 11) and, using the found A_X and X_0 values, we calculated $K_d^{(0)}$ for a cubic lattice model and a slit-like pore within the whole experimentally accessible ε_{ab} range using Eq. (3.16). The result presented in Fig. 12 shows a good agreement of the experimental data with the calculated curves. Even such a crude model as the lattice-like model and a slit-like pore can be successfully applied to assess the change in the retention volume as a function of the composition of the mobile phase.

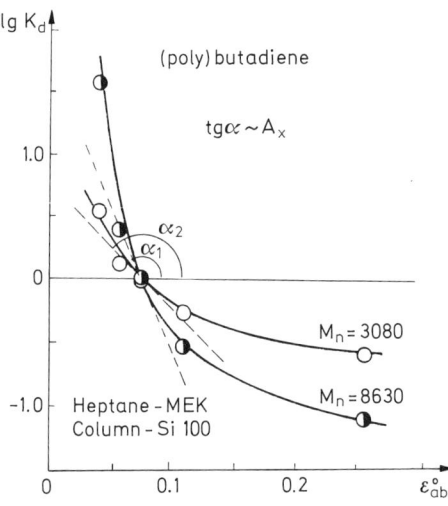

Fig. 11. The use of lg K_d vs. ε dependence close to the critical conditions to determine the parameters of energy interaction between the chain segment and the surface

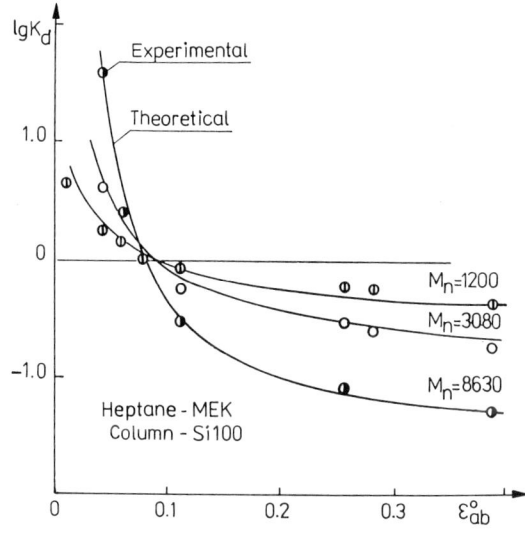

Fig. 12. Comparison of experimentally found K_d values for nonfunctional oligobutadienes with the theoretically calculated values

3.5 Universal Dependences in the Critical Region

Let us now introduce certain universal dependence for the chromatography of macromolecules close to the critical conditions. From what has been said above, it is clear that the critical conditions can be realized for different pairs of solvents. A question arises: How are the results obtained in different solvents interrelated?

Returning to the relationship of Eq. (3.15) we can write it in the form of

$$\theta = \alpha(X_0 - A_X \varepsilon_0) + \delta\theta,$$

where $\delta\theta$ describes the interactions in the mobile phase that were not taken into account before. Let us perform the following transformation

$$\theta = \alpha(X_0 - A_X \varepsilon_{ab}) + \delta\theta + \alpha A_X \varepsilon^* - \alpha A_X \varepsilon^* \approx \theta_c + A_X(\varepsilon^* - \varepsilon_{ab})$$

In Eq. (3.19), it is assumed that close to the critical conditions $\delta\theta \approx \delta\theta(C^*)$ where ε^* is the solvent power of the mixture at critical conditions. θ_c depends only on the properties of the macromolecule (stiffness [67,68]). In the absence of specific interactions, θ_c must be approximately the same in different solvents and the energy θ is a function of only the relative eluting power, $\varepsilon^* - \varepsilon_{ab}$; $\theta = \theta(\varepsilon^* - \varepsilon_{ab})$. Close to the critical conditions $K_d = K_d(N, D, \theta)$, and for a given macromolecule and adsorbent $K_d = K_d(\varepsilon^* - \varepsilon_{ab})$.

The ε^* values for different solvent pairs can differ substantially because of additional terms in Eq. (3.15), i.e. the value of ε^* is non-universal. Any slight change in the interaction of units with the surface caused either by a change in the interaction in the mobile phase or a change in the water content of the adsorbent (see Sect. 4), changes ε^*, but the form of the $K_d(\varepsilon^* - \varepsilon_{ab})$ function is approximately the same in different solvents.

If, instead of K_d, we consider $\ln K_d/M$, it can be inferred from Eq. (3.8) that at $R_0 > D$ this function should not depend on the molecular weight of the macromolecule. We can, therefore, write

$$\ln K_d/M = f(\varepsilon^* - \varepsilon_{ab}) \, . \tag{3.20}$$

The form of the function is the same for all molecules whose size is greater than the pore size D.

This dependence for polystyrene standards is shown in Fig. 13. Polybutadienes also behave in a similar way (see Sect. 4).

Let us summarize the results. In the chromatography of macromolecules similarly as in the adsorption chromatography of low-molecular-weight substances, the interaction of the units with the surface can be described in terms of a correlative approach analogous to that of Snyder. The eluting power of the binary solvent, ε_{ab}, is the main quantity characterizing the interaction of units with the surface and its variation with the composition of the mobile phase. In the case of macromolecules, however, a binary solvent has to be characterized by the relative eluting power, $\varepsilon^* - \varepsilon_{ab}$, where ε^* is the solvent power at critical conditions, when the retention volume becomes independent of molecular weight. ε^* is non-universal;

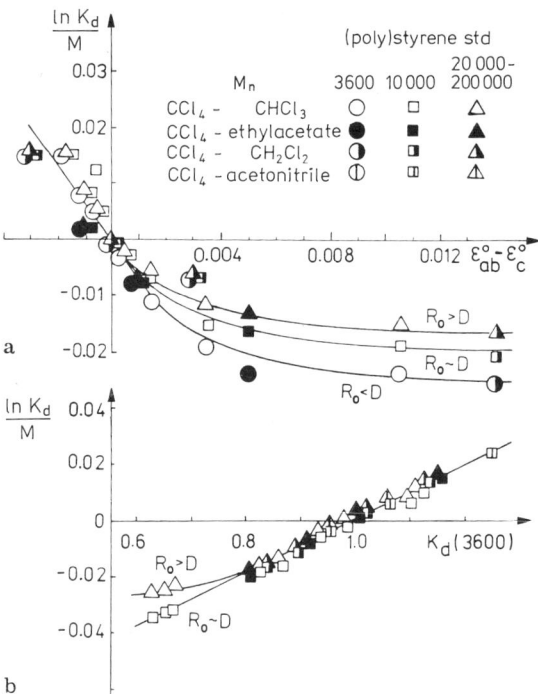

Fig. 13a and b. Universal dependence of the distribution coefficient K_d on the relative eluting power of binary solvents (**a**) and on the K_d of the polystyrene standard, $M_n = 3600$ (**b**); (Column Si-100, u = 0.5 ml/min, volume of the sample 10 µl, UV detector, λ = 275 nm, t = 27 °C)

it depends on the solvent pair and, therefore, has to be determined anew every time. After the critical conditions have been found, $K_d^{(i)}$ can be readily found either experimentally or by calculation using the simplest lattice-like model or a universal dependence of the type shown in Fig. 13.

In chromatography one traditionally avoids the use of empirical parameters, such as ε, and prefers K_d relative to a certain internal standard usually a low-molecular-weight substance. In this case, the distribution coefficients of macromolecules, K_d, are a function of $t^* - t_{ab}$, where t_{ad} and t^* are the elution times of the standard in a given mixture and in a mixture corresponding to the critical conditions. Close to the critical conditions, K_b is much more sensitive to a change in the composition of the mobile phase than to the retention times of low-molecular-weight substances, and so precision of K_d determined from $t^* - t_{ab}$ will hardly be higher than that of ε_{ab} calculated from the semiempirical Equation (3.16).

In general, it should be stated that for very long molecules the adsorption proceeds as a phase transition. Therefore, in order to work with long molecules ($M > 10^5$) at critical conditions, careful chromatographic experiments are necessary. In the case of oligomers, no such problems arise.

Strictly speaking, a change in the solvent composition changes not only the interaction of the units with the surface but also the excluded volume. Accounting for the effect of the solvent on the adsorption of macromolecules in the general case is very complicated [69]. Our experience shows that, if the components are good solvents, close to the critical conditions it seems sufficient to take into account only the change in the interaction with the surface and regard the excluded volume parameter as constant. Such an assumption provides in most cases a good estimate of the influence exerted by the composition of the mobile phase on the retention of macromolecules.

Let us now examine the application of the above concepts for analyzing the functionality type distribution of some telechelic polymers.

4 Separation of Telechelic Polymers in the Critical Region According to their Functionality Types

4.1 Examples of Separating Hydroxyl-Terminated Polyesters and Polybutadienes

Separation according to functionality at critical conditions was for the first time actualized on the hydroxyl-terminated poly(diethylene glycol adipate) (PDEGA) of $M_n \sim 2000$ using a silica gel-filled chromatographic column and hexane — methylethyl ketone (MEK) as eluent [70]

$$HO[-(CH_2)_2O(CH_2)_2OOC(CH_2)_4COO-]_n(CH_2)_2O(CH_2)_2OH$$

MEK is close to the critical solvent in the chromatography of PDEGA on silica gel. For non-functional macromolecules, there exists only a poorly resolved sepa-

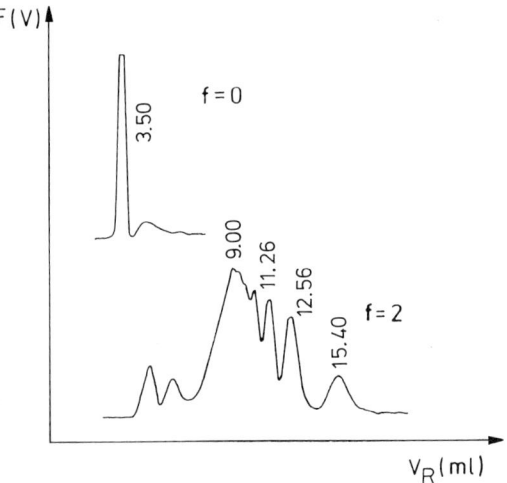

Fig. 14. Chromatograms of poly(diethylene glycol adipate) ($M_n \approx 700$) of different functionality, obtained in the exclusion separation mode. (Column Si-100, eluent methylethyl ketone, $u = 1$ ml/min, volume of the sample 10 µl, refractometer, $t = 24\ °C$)

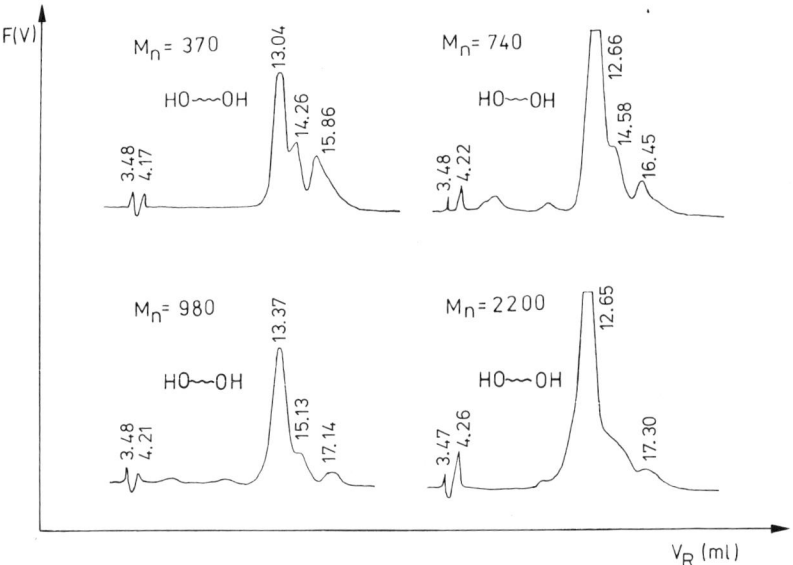

Fig. 15. FTD chromatograms of poly(diethylene glycol adipate) of different molecular weight and functionality obtained at critical conditions. (Column Si-100, eluent hexane — methylethyl ketone 8/92 vol.-%, $u = 1$ ml/min, volume of sample 10 µl, refractometer, $t = 24\ °C$)

ration according to molecular weights; for functional molecules, however, the presence of hydroxyl groups increases the resolution with respect to molecular weights, which makes separation of homologues possible (Fig. 14).

The critical conditions for strictly bifunctional PDEGA of varying molecular weight were attained by adding hexane to MEK (Fig. 15). The independence of V_R of molecular weight was observed at 8% hexane content at 24 °C.

The transition through the critical conditions has been investigated for PDEGA in the oligomer region (degree of polymerization from r = 1 to r = 10). There a derivation from the theoretical dependences for lattice-like models was expected, which is manifested by the deviation of the real V_R(lg M) calibration from the vertical line (Fig. 7) only for the functional oligomer homologues of r = 1 or r = 2.

Starting from r = 3, there is no dependence of retention volumes on the molecular weight at critical conditions.

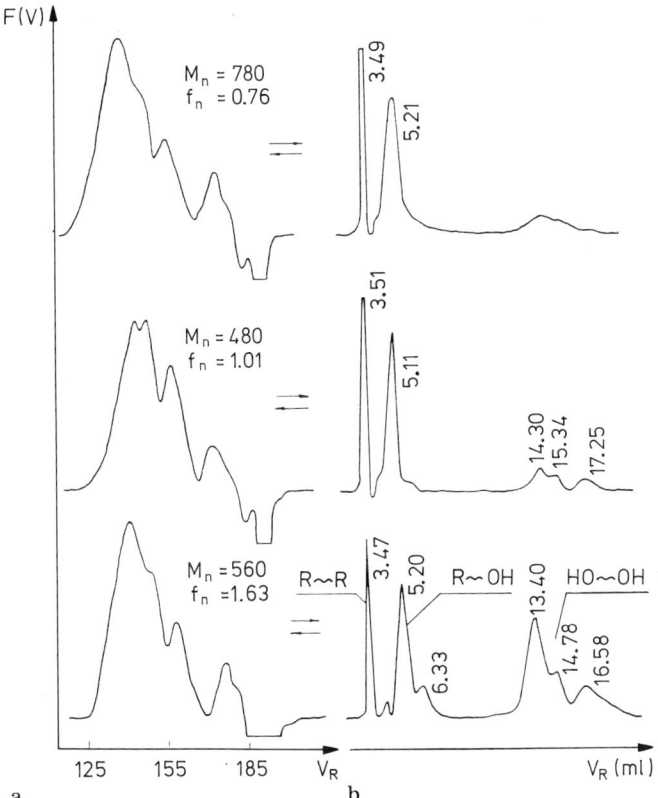

Fig. 16a and b. Gel chromatograms (a) and FTD chromatograms obtained at critical conditions (b) for poly(diethylene glycol adipates) of different molecular weight and functionality. **a** — styragel, columns 10^2, 10^3, 3×10^3, 10^4 Å, eluent tetrahydrofuran, u = 1 ml/min, volume of the sample 2 ml, refractometer, t = 24 °C;
b — column Si-100, eluent hexane-methylethyl ketone 8/92 vol.-%, u = 1 ml/min, volume of sample 10 μl, refractometer, t = 24 °C

Figure 16 shows some examples of the separation according to functionality types of industrial PDEGA samples at critical conditions. It is not difficult to select fractions corresponding to the zones of different functionality and investigate in detail their MWD distribution by standard methods in the exclusion mode.

In a similar way, one can achieve the separation according to functionality in other solvent mixtures, e.g. chloroform-acetone (Fig. 17). Except of slight differences in retention times for molecules of different functionality, which can be caused either by the difference in the interaction in the mobile phase or a change in the water content of the adsorbent, the form of FTD chromatograms for both eluents at critical conditions is similar. This indicates that for the critical conditions to be realized it is sufficient to have any two solvents suitable for the detection method, one of which works in the exclusion and the other in the adsorption mode.

Another polymer which was used to demonstrate the applicability of the proposed method was poly(butylene terephthalate) (PBTP) [71].

$$[-(CH_2)_4 OOC-\langle\bigcirc\rangle-COO-]_n$$

The sample contained zero-, mono- and bifunctional molecules:

$$CH_3-[\sim]_n-CH_3, \quad CH_3-[\sim]_n-OH, \quad HO-[\sim]_n-OH.$$

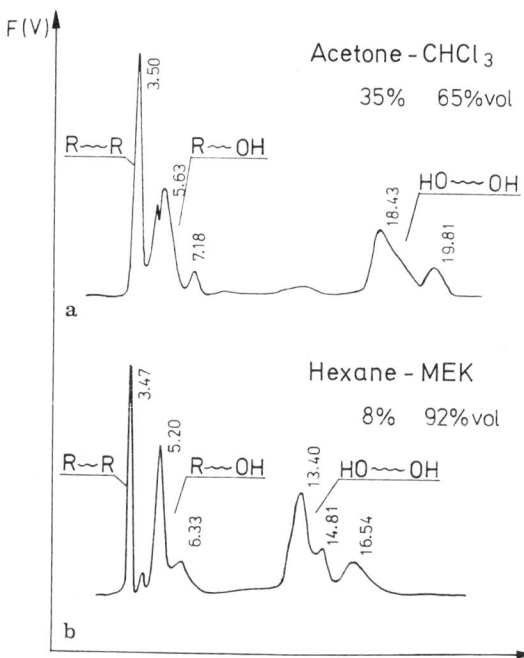

Fig. 17. FTD chromatograms of poly(diethylene glycol adipate), $M_n = 560$, $f_n = 1.63$, obtained at critical conditions in different binary solvents. (Column Si-100, u = 1 ml/min, volume of the sample 10 μl, refractometer, t = 24 °C)

Figure 18 shows chromatograms of PBTP samples, illustrating the transition through critical conditions by changing the composition of the heptane-tetrahydrofuran mixture. One can clearly see the inversion of the dependence of the retention volumes $V_R^{(i)}$ on molecular weight by passing from the adsorption to the exclusion separation mechanism and a complete independence of $V_R^{(i)}$ of the size of the molecules down to the monomer at critical conditions.

It should be particularly noted that the independence of $V_R^{(i)}$ of molecular weight at critical conditions is observed both for non-functional and functional molecules at the same eluent composition.

Knowing V_0 and V_p, one can determine $K_d^{(i)}$ from the principal chromatographic equation for mono- and bifunctional molecules at critical conditions. For the investigated system, $K_d^{(1)} = 2.2$, $K_d^{(2)} \approx 5.2$.

It is seen that $K_d^{(2)} \approx [K_d^{(1)}]^2$, and, since $\ln K_d = \Delta F$, the contribution to the free energy by the number of functional groups is additive. This feature can be explained by the fact that the size of the large macromolecule becomes in narrow pores (60 Å) comparable with the size of the pore. This is the reason for the statistical independence and additive contribution of chain ends to the free energy change ΔF.

Figure 19 shows the GPC chromatograms of three PBTP samples and the chromatograms of the same samples obtained at critical conditions. It should be noted that at critical conditions the chromatograms are identical differing only in the relative content of functional molecules, whereas their GPC chromatograms differ substantially from one another.

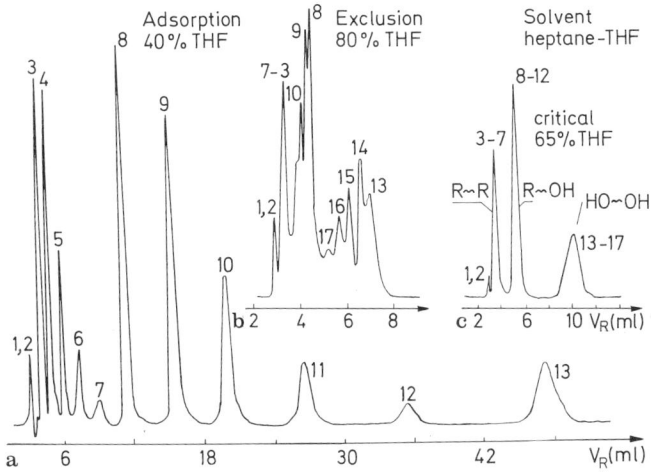

Fig. 18a–c. Transition through the critical conditions for hydroxyl-containing poly(butylene terephthalate) at a varying composition of heptane-tetrahydrofuran binary eluent. (Column Si-60, u = 1 ml/min, volume of the sample 10 µl, UV detector, $\lambda = 254$ nm, t = 24 °C)
On the chromatograms (**a**—**c**) peaks 1, 2 — solvent; peaks 3–7 nonfunctional, homologues f = 0, with degree of polymerization r = 0–4, M = 90 + 228r; peaks 8–12 (f = 1, r = 0–4); peaks 13–17 (f = 2, r = 0–4)

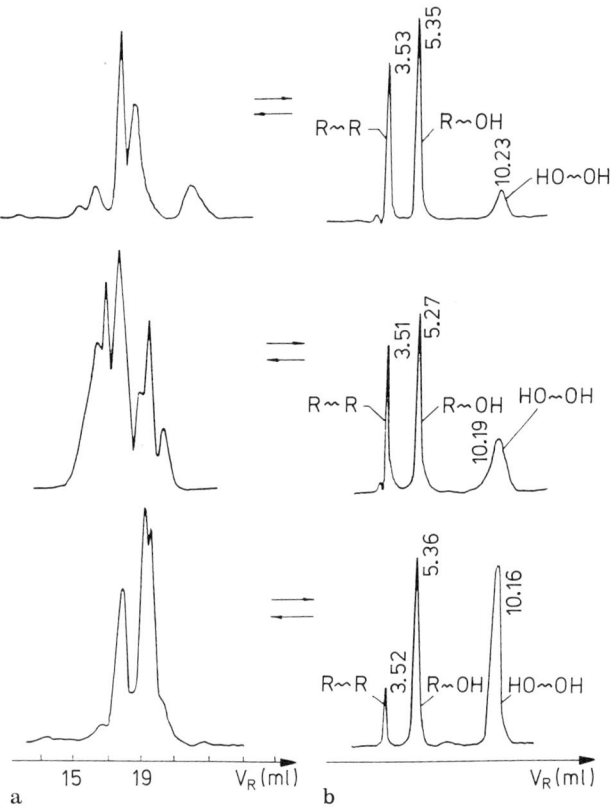

Fig. 19a and b. Gel chromatograms (a) and FTD chromatograms in critical conditions (b) for poly-(butylene terephthalate) of different molecular weight and functionality. a — microstyragel, columns 50, 100, 500 Å, eluent chloroform, u = 1 ml/min, volume of sample 10 μl, UV detector, λ = 254 nm; b — column Si-60, eluent heptane-tetrahydrofuran 35/65 vol.-%

As in the previous case, one can try to find the critical conditions and determine FTD in another pair of solvents, and take, e.g. chloroform as a chromatographically strong solvent so that the exclusion mode becomes operative for PBTF (Fig. 20). In this case, however, $K_d^{(i)}$ for mono- and bifunctional molecules is much higher than for the heptane — THF system ($K_d^{(1)} \sim 10$, $K_d^{(2)} \sim 100$). Therefore, the given pair is unsuitable for FTD determination using the chosen column as bifunctional molecules are practically not eluted. This example demostrates the role of the interactions in the mobile phase. In the previous case, the strong interaction of THF molecules with the OH groups of the polymer decreases the interaction energy with the surface [cf. Eq. (3.14)]. Thus it is possible to determine FTD at critical conditions in the mixture heptane — THF. To decrease K_d for the heptane — $CHCl_3$ system, one can choose an adsorbent with a larger pore size [cf. Eq. (3.13)] or use the gradient to elute bifunctional molecules.

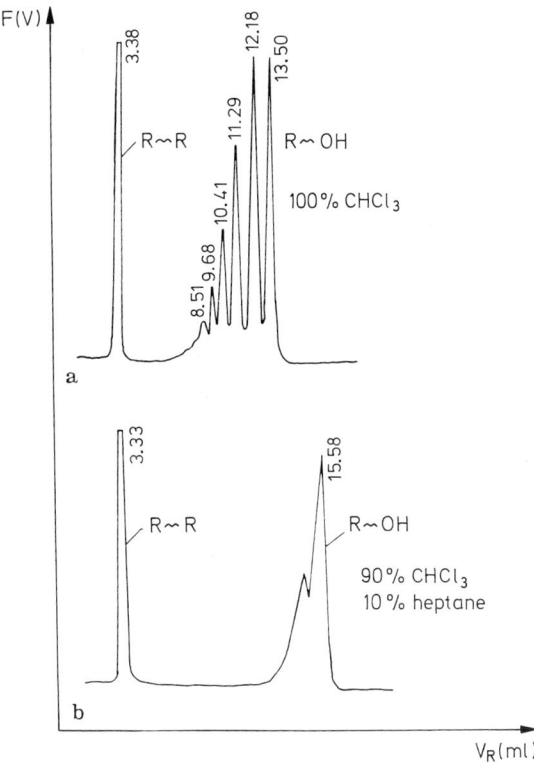

Fig. 20a and b. Separation of poly-(butylene terephthalate) in heptane-chloroform binary eluent at critical conditions (**b**) and close to critical condition (**a**). (Column Si-60, u = 1 ml/min, volume of the sample 10 μl, UV detector, λ = 254 nm, t = 24 °C)

An analysis of the expressions for $K_d^{(i)}$ at critical conditions [Eq. (3.13)] shows that for the chosen adsorbent the scope of retention volumes is determined by the $\sigma = \exp(\theta_f - \theta_c)$ factor, i.e. it grows exponentially with increasing difference between the interaction energies of the functional groups and the groups in the chain. If this difference is large, also $K_d^{(i)}$ for the functional molecules can become very large.

This is the case, e.g. for oligobutadienes with terminal OH groups. Their behaviour in the chromatography on silica gel is analyzed below.

The transition through critical conditions for polybutadienes already has been considered (Fig. 6). The critical conditions corresponded to 0.5 vol.-% MEK in heptane. The solvent power calculated from Eq. (3.16) is given by $\varepsilon^* \approx 0.1$. For the used adsorbent with D = 100 Å and $a \approx 10$ Å, the retention volumes of functional molecules were estimated from Eq. (3.13). Using the tabular values of X(—OH) = 5.60 and A_X(—OH) = 8.5 [64], and taking for the terminal segment $(X_0)_f = 5.60$ and $(A_X)_f = 10$, we find $\theta_f = 6.66$ at critical conditions. From Eq. (3.13) we get $K_d^{(1)} \sim 10^2$, $K_d^{(2)} \sim 10^4$.

For the investigated oligobutadiene (OBD) — silica gel system at critical conditions, the retention volumes $V_R^{(1)}$ and $V_R^{(2)}$ are thus too high, and mono- and bifunctional molecules are not eluted from the column.

For OBD functional molecules, the $K_d^{(1)}$ and $K_d^{(2)}$ values at critical conditions can be decreased by increasing the pore size. However, for such a great difference in the adsorption energies of the chain segment θ and the terminal segment θ_f characteristic for oligobutadienes measurable distribution coefficients cannot be obtained even if D increased up to 10^3 Å (Fig. 21). That is why one has to go deeply into the exclusion region, in order to perform separation according to functionality in the isocratic mode and using compositions of a higher eluting power ($\varepsilon_{ab} > \varepsilon^*$).

For the 15% MEK — 85% heptane mixture, $\varepsilon_{ab} = 0.316$. Using the redetermined values of $X_0 = 0.2$ and $A_X = 0.82$ for the chain segment and taking $(X_0)_f = 5.6$ and $(A_X)_f = 10$ for the terminal segment containing the functional group, one can calculate the theoretical calibration $V_R = V_R(\lg M)$ and FTD chromatograms for mono- and bifunctional molecules of oligobutadienes [66].

Theoretical calibrations for mono- and bifunctional molecules were described by equations $\lg M_n = 5.05 - 0.34 V_R$, $\lg M_n = 4.6 - 0.12 V_R$. When FTD chromatograms were calculated, it was assumed that the MWD fraction $P_r^{(i)}$ (r is the degree of polymerization) for the polymer of functionality i is approximated by the expression

$$P_r = P_0 \frac{\gamma^\omega r^\omega}{(\omega - 1)!} e^{-\gamma r},$$

where γ and ω are parameters, $r_n = \omega/\gamma$, $r_w/r_n = (\omega + 1)/\omega$, $M_n = rM_{un}$, where $M_{un} = 72$ is the molecular weight of the oligobutadiene unit.

The P_r function is easily transformed into $P(V)$, which determines the $F(V)$ chromatogram without taking into account the instrumental broadening, by the following expression: $F(V) = P(V) B$, ($\lg M = A - BV_R$).

FTD chromatograms calculated in this way for a series of samples (Fig. 22) are similar to the experimental ones and thus an unambiguous separation of oligomers according to their molecular weight and functionality is possible.

The theoretical V_R vs. $\lg M$ dependences for macromolecules of differing functionality have been compared with those obtained experimentally in Fig. 23 (the points indicate the experimental results).

It is seen from this figure that in the exclusion mode the presence of the functional groups increases the resolution of separation according to the molecular weights because the functional groups are more strongly adsorbed than the backbone units. Thus the MWD even for very narrow polybutadiene samples ($M_w/M_n = 1.05$) can be investigated in detail.

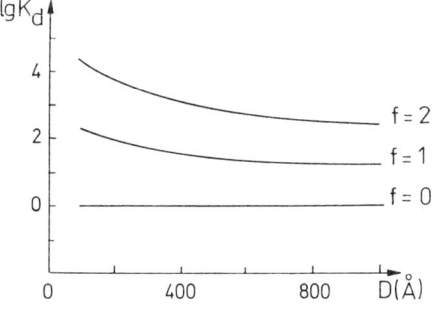

Fig. 21. Distribution coefficients of hydroxyl-containing mono- and bifunctional oligobutadienes at critical conditions for different pore sizes of the adsorbent

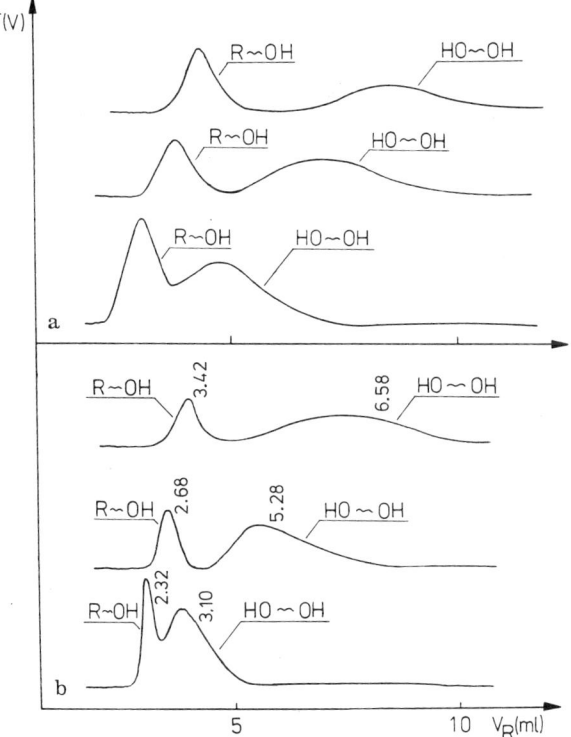

Fig. 22a and b. Theoretical (**a**) and experimental (**b**) FTD chromatograms of oligobutadienes with terminal OH groups. (Column Si-100, eluent heptane-methylethyl ketone 85/15 vol.-%, u = 1 mg/min, volume of sample 10 μl, refractometer, t = 24 °C)

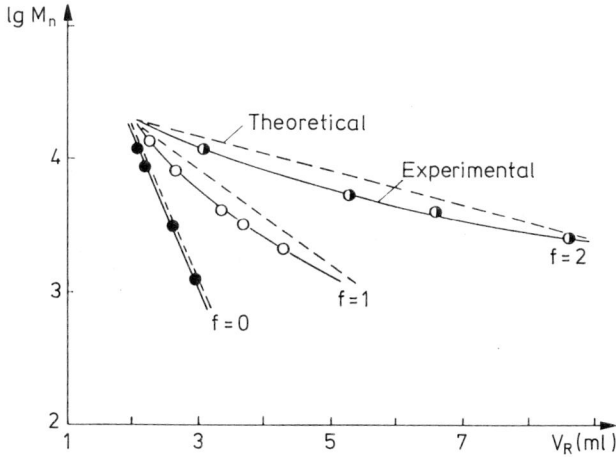

Fig. 23. Calibration dependences for hydroxyl-containing oligobutadienes of different functionality. (Column Si-100, eluent heptane-methylethyl ketone 85/15 vol.-%, u = 1 ml/min, volume of the sample 10 μl, refractometer, t = 24 °C)

However, if the samples had a wider MWD, the overlapping of the zones of differing functionality would be much stronger. This overlapping is not connected with the efficiency of the separation system (in our case the column had ~ 4000 theoretical plates). In such a situation it would be more convenient to use a gradient, but in working with OBD problems of detection arise.

It is also possible to examine another stationary phase, with a smaller difference in the interaction energies of the terminal and backbone units, e.g. a reversed phase.

Another way of decreasing the retention volume of functional macromolecules can be found in raising the column temperature. In this case, it should be borne in mind that a change in temperature causes a number of changes in the chromatographic system, mainly associated with the shift in the equilibrium between the components of the adsorbed mobile phase and water, which results in a change in the ε_{ab} of the mixture and the adsorbent activity.

4.2 Some Methodological Problems Concerning the Chromatography of Macromolecules in the Critical Region

If separation according to functionality is performed in the critical region, problems sometimes arise associated with a very high or a very small difference in the interaction energy of the terminal and chain units. If the difference is small, it is necessary to use a more selective stationary phase or to increase the separation efficiency. If the difference is large, one should increase the pore size or use gradient elution in order to achieve separation according to functionality and molecular weights.

Qualitatively, the influence of the eluent gradient on chromatographic separation according to functionality is shown in Fig. 24. The initial point on the ε_{ab} axis roughly determines the absolute values of retention volumes at which the macromolecules are eluted, and the slope of ε_{ab} vs. V curve the distance between the zones of different functionality. Unpublished experimental data obtained in gradient chromatography of PBTP in a binary heptane — tetrahydrofuran eluent fully support this conclusion.

When one uses gradient elution, both in the adsorption and the exclusion separation modes, it is necessary to bear in mind that the zones of different functionality may overlap which is caused by the MWD, since in both modes K_d very strongly (exponentially) depends on the length of the molecule, while the functionality dependence is given by Eq. (3.11). Moreover, gradient elution markedly limits the choice of a detector and is, therefore, applicable only for some polymers.

The next problem to be examined deals with the effect of water contained in the adsorbent on the chromatographic behaviour of macromolecules. The role of water in adsorption chromatography on polar adsorbents such as silica gel is extremely important [63, 64], because its varying content in the adsorbent often yields poorly reproducible results.

Figure 25 shows a transition through critical conditions of polybutadienes in two solvent pairs. After separation, using substances with known X_0 in hexane ($\varepsilon_0 = 0$) and a standard procedure [64], one can determine the water content in the column

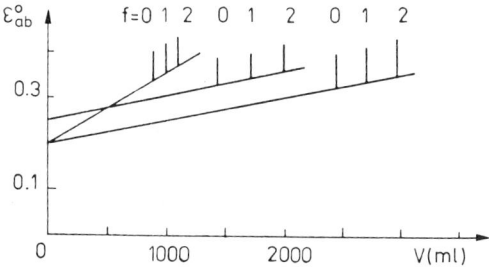

Fig. 24. Examples of the effect of eluent programme $\varepsilon_{ab}(V)$ on retention volumes of macromolecules of different functionality (calculated data for $D = 8, N = 10, X_0 = 3.6, X_{0f} = 5.6, V_0 = V_p = 40$ ml, verical lines in the figure correspond to retention volumes of macromolecules of different functionality)

(Fig. 26). It is seen that in the second case the water content is much lower, which is documented by a strong difference in ε^*.

Additional terms have to be inserted into Eq. (3.15) to account for the water content in the adsorbent relative to the standard absolutely dry adsorbent [72)]

$$\theta = \alpha(X_0 - A_x \varepsilon_0) + \Delta\theta$$

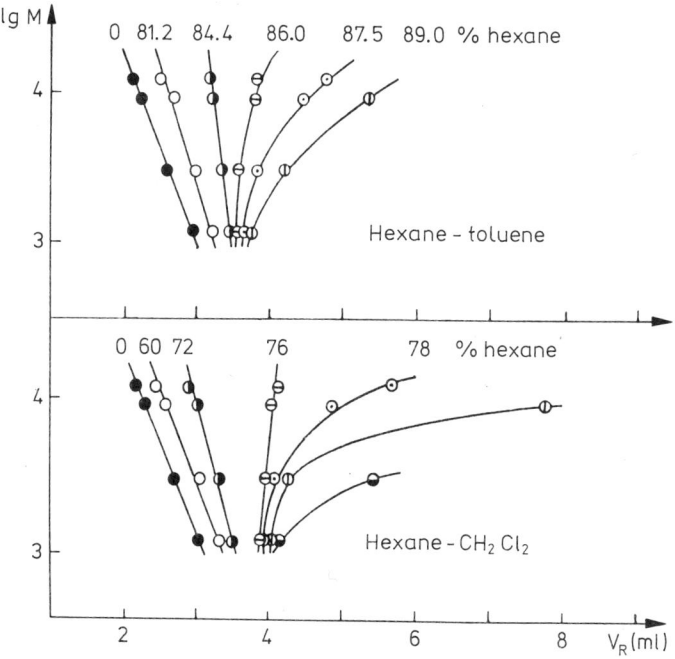

Fig. 25. Transition through the critical conditions for nonfunctional oligobutadienes in different binary solvents. (Column Si-100, $u = 1$ ml/min, volume of the sample 10 μl, refractometer, $t = 24$ °C)

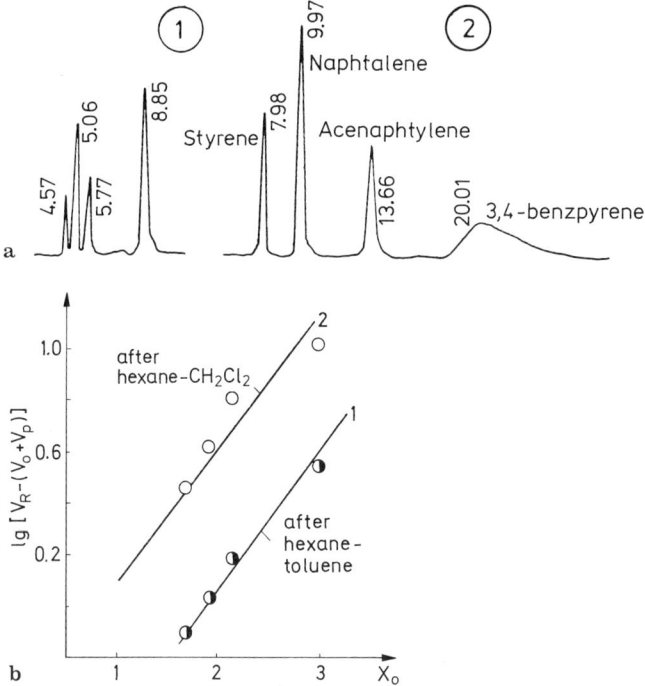

Fig. 26a and b. The effect of water content in the adsorbent on the retention volumes of standards in hexane (**a**) and determination of adsorbent activity α and monolayer volume (**b**) after reaching the critical conditions in binary solvents given in Fig. 25. X_0 = 1.71 styrene, 2.02 naphthalene, 2.14 acenaphthylene, 3.02 benzopyrene-3,4 [64]

However, a change in the water content of the adsorbent after the solvent has been replaced is sufficiently slow, and $\Delta\theta$ can be regarded as changing only slightly during the course of the experiment performed close to critical conditions. This assumption makes possible to express θ as a function of $\varepsilon^* - \varepsilon_{ab}$, thus excluding $\Delta\theta$. The ln K_d/M vs. $\varepsilon^* - \varepsilon_{ab}$ dependence is shown in Fig. 27.

The water content in the column can change from day to day, even if one and the same solvent mixture is used. To ensure good reproducibility when working at critical conditions it is, therefore, necessary to take special measures to control the adsorbent humidity.

Let us now formulate the general scheme of choosing the chromatographic system and the separation conditions for the analysis of functionality type distribution.

1. For the investigated type of telechelic polymer (preferably a sample containing only nonfunctional macromolecules so that retention volumes are close to $V_0 + V_p$), it is first of all necessary to choose two solvents, in one of which (polar solvent b with ε_{0b}) the exclusion mode is operative and in the other (nonpolar solvent a with ε_{0a}) the adsorption mode. The solvents must be able to dissolve readily the analyzed samples and satisfy the detection conditions.

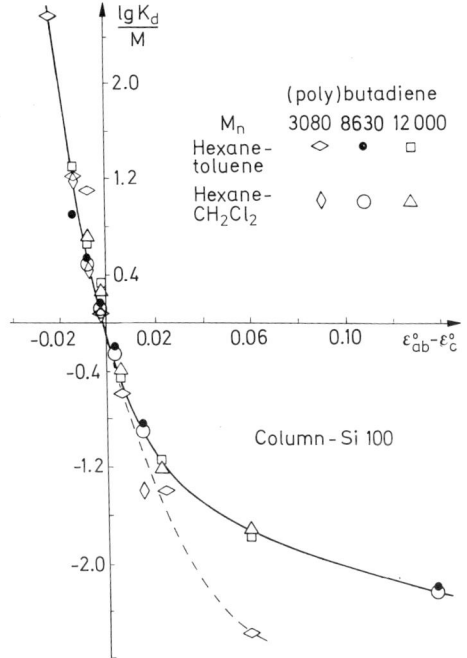

Fig. 27. The dependence of ln K_d/M for polybutadienes of different molecular weight on the relative elution power of binary solvents

By varying the composition of these solvents, the critical conditions and the corresponding values of C* and ε* are found. For this purpose, it is necessary to have 4–6 samples (in the last resort two will suffice) of different molecular weight. Pore size D, must be of the same order as the size of the macromolecules.

2. The functional molecules are then separated at the critical conditions. Before measurements, it is useful to estimate the scale of retention volumes for the functional molecules ($K_d \sim 10$ is desirable) using Eq. (3.13) and the tabular adsorption energy values for chain groups, or passing through the column at critical conditions the low-molecular-weight analogues (or model samples) of different functionality.

3. If the separation at critical conditions is impossible because of very high retention volumes of the functional macromolecules, it is necessary to take an adsorbent with wider pores or proceed to the exclusion region ($\varepsilon_{ab} > \varepsilon^*$) and perform the chromatography in the isocratic mode, or to use gradient elution. If this is not successful, another adsorbent has to be chosen which is less selective with respect to functional groups.

4. If separation according to functionality cannot be achieved because of a small difference between the interaction energies of the functional group and the chain groups, one can try to improve the efficiency of the separation system, but it is necessary to choose another stationary phase which is more selective with respect to functional groups.

In our opinion, the routes 1–4 provide the optimal approach to choosing a chromatographic system and conditions for the separation of telechelic oligomers according to their functionality types. The critical conditions are usually found within a sufficiently short time. If, however, the separation of a specific oligomer on a specific adsorbent is not good, certain difficulties arise. However, the experience in modifying the stationary and the mobile phases in liquid chromatography allows one to solve successfully this problem as well, since the practically important groups are not so numerous, and it is always possible to select the appropriate separation conditions.

The experimental data described above will now be commented on with respect to the theoretical concepts developed in Section 3.

The first important question to be answered concerns the lower limit of molecular weight down to which the concepts obtained for long Gaussian or "lattice-like" chains were applicable. It is clear that a Gaussian chain does not adequately describe the conformational properties of short oligomer chains. Other models e.g. a model of the wormlike chain may be more suitable. The introduction of this model may lead to considerable mathematical complications and the determination of K_d may become difficult.

Fortunately, the concepts developed for long chains are also perfectly suitable in the oligomer region. The critical conditions at which there is no dependence of retention volumes on the size of the molecules extends for oligomers down to very low molecular weights. For PDEGA, only bifunctional homologues with a degree of polymerization of $r = 1, 2$ fall out of the general V_R vs. lg M dependence, but then conventional methods of low-molecular weight chromatography can be applied. For PBTP, the general retention volume vs. molecular-weight dependence holds down to the monomer.

In both examples, we have the case of $R_0 < D$. Eq. (3.13), however, are suitable for a good estimation for the retention volumes at critical conditions even if $R_0 < D$.

For polybutadienes of $M_n \approx 10^4$, the lattice-like model already gives a good description of their conformational properties (each segment is composed of about two monomers). The errors introduced by the lattice-like model appear to be insignificant.

It is possible to design a theory for a chain consisting of several freely jointed rods [73]. Such a model reveals certain non-monotonous dependences of K_d on the number of rods (molecular weight) close to the critical conditions.

Let us now turn again to the problem of the thermodynamic quality of the solvent. In the above examples, the solvent quality was different. Thus, for PDEGA in the hexane — MEK mixture, only MEK is good; on the contrary, for polybutadienes hexane is a good solvent. In the chloroform-acetone mixture both solvents are good for PDEGA. Thus, the use of solvents with a different thermodynamic quality in binary mixtures has no qualitative effect on separation; it is only necessary that the sample should be readily soluble in binary mixtures whose composition is close to C*.

Let us recollect Fig. 10 showing the dependence between the eluting power of a binary solvent, ε_{ab} and the molar fraction of the polar component, C_b. Two characteristic parts can be seen on the curve: a rapid increase in ε_{ab} at small C_b and a plateau at larger C_b. In the plateau region one can very accurately fix the critical conditions, since there ε_{ab} depends but very slightly on C_b. This is the case for PDEGA and it is necessary that ε_{0b} is a little greater than ε^*.

In the plateau region, a small change in ε_{ab} requires a large change in the composition of the binary mixture, which may greatly change the excluded volume parameter. To reduce this influence it seems more convenient to work in the region of increasing ε_{ab}, since here the transition from the exclusion to the adsorption mode occurs a very narrow region of ΔC_b.

The use of Eqs. (3.13) for the determination of distribution coefficients is not always possible, because of the unknown correlation parameters, e.g. for the reversed phase. It is much more convenient to determine $K_d^{(i)}$ of functional molecules by examining at critical conditions a low-molecular-weight substance of a similar chemical structure and containing the same functional groups. Suitable for this purpose is a low-molecular-weight monofunctional homologue (the assumption $K_d^{(2)} \approx [K_d^{(1)}]^2$ is sufficient for practical applications).

Above we have considered the examples of separation according to functionality on silica gel as the stationary phase. Silica gel has been prefered for certain functional groups and because of the existence of a well-founded correlation theory for silica gel [64]. At present, however, most separations are performed on chemically modified phases. All the basic regularities of the chromatography of macromolecules close to the critical conditions are also valid for these stationary phases.

5 Method for Analyzing Other Types of Macromolecular Heterogeneity

In the previous sections, we have dealt only with telechelic oligomers containing linear non-, mono- and bifunctional molecules. In practice, however, oligomer systems containing components of higher functionality are often encountered. If the mean square distance between functional groups in the chain is greater than the pore size of the stationary phase, then

$$K_d^{(n)} \approx [K_d^{(1)}]^n .$$

At critical conditions this relationship can be used at any ratio of pore size to the distance between functional groups.

Branched polyfunctional molecules of star-like and comb-like types are often encountered and, therefore, it is necessary to examine the effect of branching on the chromatographic behaviour of a macromolecule near the critical point.

If the chemical structure of the branch point is similar to that of the chain unit, the behaviour of branched molecules is determined rather by the functional groups than by branching.

Let us now consider the behaviour of cyclic macromolecules at critical conditions. The macrocycle at critical conditions be transferred from the liquid phase into the pore in the following way: (a) transform a cycle of length N into a linear molecule ($\Delta F = -3/2 \ln N$); (b) transfer the linear molecule into the pore; since this occurs at critical conditions, $\Delta F = 0$; (c) form again a cycle in a slit-like (two-dimensional) pore ($\Delta F = \ln N$). In going from (a) to (c), it is seen that the total free energy change for the transfer of the macrocycle into the pore at critical conditions is not equal to zero but depends on the size of the ring

$$\Delta F = -1/2 \ln N .$$

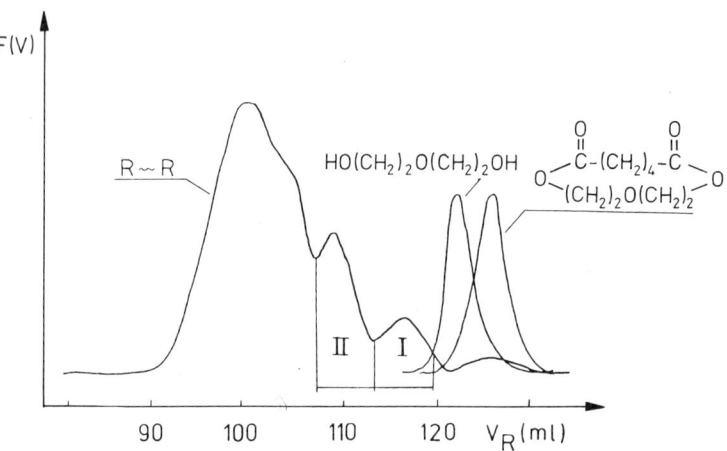

Fig. 28. Gel chromatograms of poly(diethylene glycol adipate) with terminal butoxyl groups, $M_n = 700$, $f = 0$, diethylene glycol and a ring consisting of adipic acid and diethylene glycol, $M = 316$ styragel columns 60, 10^3, 10^4 Å, eluent tetrahydrofuran, $u = 1$ ml/min, volume of the sample 2 ml, refractometer, $t = 24\ °C$

At critical conditions, it is thus more advantageous for the macrocycle to be located inside the pore, and, therefore, linear and cyclic macromolecules will be separated under these conditions in the whole range of their size and MWD. Data on the separation of linear and cyclic PDEGA are presented in Figs. 28 and 29 [70]. Fractions of different molecular weight were taken from a sample of a non-functional oligomer containing linear and cyclic macromolecules, Fig. 28. At critical conditions, each fraction was "split" into two — the linear and the cyclic one. As expected, no dependence of retention volumes on the molecular weight was found for linear molecules, whereas for cyclic molecules the retention volume increases with increasing molecular weight (Fig. 29).

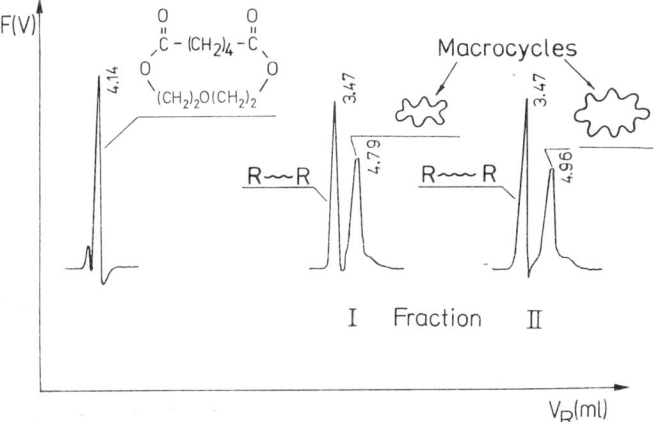

Fig. 29. Chromatograms of the fractions of poly(diethylene glycol adipate) and cyclic molecules obtained at critical conditions. (Column Si-100, eluent hexane-methylethyl ketone 8/92 vol.-%, $u = 1$ ml/min, volume of sample 10 μl, refractometer, $t = 24\ °C$)

It should be noted that the critical condition approach can also be used for the analysis of block copolymers. At present, there exists a theory based on the lattice-like model [74] describing different modes of chromatography of block copolymers which makes possible to find conditions for separation according to composition. The analysis of experimental work pertaining to this question is, however, beyond the scope of this review.

In conclusion, it should be said that close to the critical conditions in a weak adsorption mode there exists a strong dependence of retention volume V_R on concentration of the solute. Chromatography close to the critical conditions can, therefore, be used as a convenient method for investigating the adsorption isotherms in a weak adsorption mode, which is of the greatest theoretical interest [75] and has not yet been studied experimentally.

6 Conclusions

In this review the primary emphasis has been placed on the problem of determining the FTD of telechelic polymers; the mechanisms of reactions leading to functional defectiveness have not been discussed. However, for solving this problem the FTD and MWD of a polymer have to be investigated. Even in the simplest cases considered above, the chromatographic determination of FTD and MWD is not simple.

We have tried to explain the general principles of the chromatography at critical conditions applicable not only to functionality determination. In order to analyse the herterogeneities in macromolecules, it is necessary to take into account not only the relationships between the macromolecule size and the pore size of the stationary phase, but also the energy of the interaction of the molecule with the adsorbent.

Rather much attention has been paid to purely chromatographic problems. This emphasis seems appropriate in the present review since the problems of determining the FTD of macromolecules reduces to conventional problems of low-molecular-weight adsorption chromatography. Moreover, this approach provides a basis for investigating other more complex types of macromolecular heterogeneity by means of high-performance liquid chromatography.

Acknowledgements: The authors wish to express their gratitude to A. M. Skvortsov, A. A. Gorbunov and T. M. Birshtein for helpful discussion, to S. M. Baturin, Ya. I. Estrin and A. I. Kuzayev, who kindly provided a series of hydroxyl-terminated oligomers, as well as to V. V. Gurianova and T. N. Prudskova for carrying out some of the experiments.

7 References

1. Entelis, S. G., Evreinov, V. V., Kuzaev, A. A.: Reactive Oligomers. Functionality Type Distribution, in: Uspekhi khimii i fiziki polimerov (Progress in Polymer Chemistry and Physics) (ed.) Rogovin, Z. A., p. 201, Moscow, Khimiya, 1973

2. Ol'khov, Yu. A. et al.: Vysokomol. soedin. *A14*, 2662 (1972)
3. Strecker, A. H., French, D. M.: J. Appl. Polym. Sci., *12*, 1967 (1968)
4. Valuyev, V. I. et al.: Vysokomol. soedin. *A14*, 11 (1972)
5. Consage, J. P.: J. Appl. Polym. Sci. *14*, 2157 (1970)
5a. Matejka, L., Dušek, K.: Polymer Bulletin., *3*, 489 (1980)
6. French, D. M., Strecker, A. H.: J. Macromol. Sci. *A2*, 893 (1971)
7. Evreinov, V. V. et al.: Vysokomol. soedin. *A20*, 2146 (1978)
8. Fogiel, A. W.: Macromolecules *2*, 581 (1969)
9. Stockmayer, W. H.: J. Polym. Sci. *9*, 69 (1952)
10. DiMilo, A. J., Jonson, D. E.: J. Macromol. Sci. *A3*, 1419 (1969)
11. Evreinov, V. V. et al.: Vysokomol. soedin. *A12*, 829 (1970)
12. Muenker, A. H., Hudson, B. E.: J. Macromol. Sci. *A3*, 1465 (1969)
13. Low, R. D.: J. Polym. Sci. *A9*, 589 (1971)
14. Shlyakhter, R. A., Valuev, V. I.: Investigation of the Functionality of Oligomers, in: Gel'-pronikayushchaya khromatografiya (Gel Permeation Chromatography) (ed.) Kuzayev, A. I., p. 73, Chernogolovka, Inst. of Chem. Phys. USSR Acad. of Sciences 1974
15. Valuev, V. I. et al.: Vysokomol. soedin. *B19*, 172 (1977)
16. Valuev, V. I. et al.: Ibid. *A15*, 2728 (1973)
17. Valuev, V. I. et al.: Application of Liquid Adsorption Chromatography for the Analysis of Liquid Rubbers according to the Functionality Types. in: Metody analiza i kontrolya kachestva produktsii v khimicheskoi promyshlennosti (Methods of Analysis and Quality Control in Chemical Industry), p. 24, Moscow, NIITEKhIM, 1978
17a. Pokorný, S. et al.: J. Liquid Chromatog., 4(1), 1 (1981)
18. Tsvetkovsky, I. B. et al.: Vysokomol. soedin. *B19*, 645 (1977)
19. Valuev, V. I. et al.: Zh. anal. khim. *30*, 1235 (1975)
20. Tsvetkovsky, I. B. et al.: Vysokomol. soedin. *A17*, 2609 (1975)
21. Kuzaev, A. I., Entelis, S. G.: Adsorption of Polyethers on Silica Gel Surface, in: Makromolekuly na granitse razdela faz (Macromolecules at the Interphase), p. 92, Kiev, Naukova dumka, 1971
22. Radugin, V. S. et al.: Column Chromatography (ed.) Kuzayev, A. I., p. 82, Chernogolovka, Inst. of Chem. Phys. USSR Acad. of Sciences 1974
23. Kuzayev, A. I., Suslova, Ye. N.: Vysokomol. soedin. *A15*, 1178 (1973)
24. Kuzayev, A. I.: Possibilities of Adsorption Chromatography in Polymer Analysis, in: Gel'-pronikayushchaya khromatografiya (Gel Permeating Chromatography) (ed.) Kuzayev, A. I., p. 35, Chernogolovka, Inst. of Chem. Phys. USSR Acad. of Sciences 1974
25. Vakhtina, I. A. et al.: Vysokomol. soedin. *A20*, 2388 (1978)
26. Vakhtina, I. A., Khrenova, R. I., Tarakanov, O. G.: Zh. anal. khim. *28*, 1625 (1973)
27. Vakhtina, I. A., Tarakanov, O. G., Khrenova, R. I.: Vysokomol. soedin. *A16*, 2598 (1974)
28. Vakhtina, I. A. et al.: Plast. massy *4*, 56 (1976)
29. Vakhtina, I. A., Tarakanov, O. G.: Plaste Kautschuk *21*, N1, 28 (1974)
30. Belenkii, B. G., Vakhtina, I. A., Tarakanov, O. G.: Vysokomol. soedin. *A17*, 2116 (1975)
31. Belenkii, B. G. et al.: J. Chromatogr. *129*, 115 (1976)
32. Vakhtina, I. A., Tarakanov, O. G.: Plaste Kautschuk *6*, 401 (1976)
33. Vakhtina, I. A. et al.: Vysokomol. soedin. *A22*, 1671 (1980)
34. Vakhtina, I. A. et al.: A Study of the Functionality and Molecular Weight Dispersity for Stiff Foams, in: Sintez i fiziko-khimiya polimerov (Synthesis and Physico-Chemistry of Polymers), p. 58, Kiev, Naukova dumka, 1976
35. Knopp, Kh. et al.: Vysokomol. soedin. *A22*, 1788 (1980)
36. Kuzaev, A. I. et al.: Zh. fiz. khim. *42*, 1235 (1974)
37. Evereinov, V. V., Eremeeva, T. V., Sarynina, L. I.: Vysokomol. soedin. *A16*, 1884 (1974)
38. Filatova, N. N. et al.: Ibid. *A20*, 2367 (1978)
39. Tyutyundzhan, I. P. et al.: Ibid. *A17*, 104 (1975)
40. Filatova, N. N. et al.: Plast. massay *4*, 40 (1979)
41. Vakhtina, I. A. et al.: Vysokomol. soedin. *A18*, 471 (1976)
42. Vakhtina, I. A. et al.: Ibid. *A23*, 1096 (1981)
43. Dmitrieva, T. S. et al.: Kauchuk i rezina *10*, 6 (1977)
44. Kuzaev, A. I.: Vysokomol. soedin. *A22*, 2082 (1980)

45. Nasonova, T. P. et al.: Ibid. *A13*, 635 (1971)
46. Casassa, E. F.: J. Polym. Sci. *B5*, 773 (1967)
47. Casassa, E. F., Tagami, Y.: Macromolecules *2*, 14 (1969)
48. DiMarzio, E. A., Rubin, R. J.: J. Chem. Phys. *55*, 4318 (1971)
49. Skvortsov, A. M. et al.: Vysokomol. soedin. *A20*, 816 (1978)
50. DeGennes, P.-G.: Scaling Concepts in Polymer Physics. Ithaca London, Cornell University Press, 1979
51. Skvortsov, A. M. et al.: Vysokomol. soedin. *A20*, 678 (1978)
52. Nefedov, P. P., Lavrenko, P. N.: Transport Methods in the Analytical Chemistry of Polymers, 232 p., Leningrad, Khimiya, 1979 (in Russian)
53. Lifshits, I. M.: Zh. eksp. teor. fiziki *55*, 2408 (1968)
54. Lifshits, I. M., Grosberg, A. Yu., Khokhlov, A. R.: Rev. Mod. Phys. *50*, 683 (1978)
55. Lifshits, I. M., Grosberg, A. Yu., Khokhlov, A. R.: Uspekhi fiz. nauk *127*, 353 (1979)
56. De Gennes, P. G.: Rept. Progr. Phys. *32*, 187 (1969)
57. Tennikov, M. B. et al.: Vysokomol. soedin. *B19*, 657 (1977)
58. Belenkii, B. G. et al.: J. Chromatogr. *147*, 99 (1978)
59. Skvortsov, A. M., Gorbunov, A. A.: Vysokomol. soedin. *A22*, 2641 (1980)
60. Gorshkov, A. V., Evereinov, V. V., Entelis, S. G.: Ibid. *A24*, 524 (1982)
61. Skvortsov, A. M., Gorbunov, A. A.: Ibid. *A22*, 1137 (1980)
62. Skvortsov, A. M., Zhulina, Ye. B., Gorbunov, A. A.: Ibid. *A22*, 820 (1980)
63. Engelhardt, H.: Hochdruck-Flüssigkeits-Chromatographie. Berlin—Heidelberg—New York, Springer, 1977
64. Snyder, L. R.: Principles of Adsorption Chromatography. New York, Dekker, 1968
65. Gorshkov, A. V. et al.: Zh. fiz. khim. *56*, 2641 (1982)
66. Gorshkov, A. V., Evreinov, V. V., Entelis, S. G.: Ibid. *57*, 3 (1983)
67. Skvortsov, A. M., Birshtein, T. M., Zhulina, Ye. V.: Vysokomol. soedin. *A18*, 1993 (1976)
68. Birshtein, T. M., Zhulina, E. V., Skvortsov, A. M.: Biopolymers, *18*, 1171 (1979)
69. DeGennes, P. G.: J. Phys. *37*, 1445 (1976)
70. Gorshkov, A. V., Evreinov, V. V., Entelis, S. G.: Dokl. AN SSSR, *273* (1983)
71. Gorshkov, A. V. et al.: Vysokomol. soedin. *B27*, 181 (1985)
72. Gorshkov, A. V., Evreinov, V. V., Entelis, S. G.: Zh. fiz. khim., *59*, 1475 (1985)
73. Zhulina, Ye. V., Gorbunov, A. A., Skvortsov, A. M.: Vysokomol. soedin., *A26*, 915 (1984)
74. Skvortsov, A. M., Gorbunov, A. A.: Vysokomolek. soyed. *A21*, 339 (1979)
75. DeGennes, P. G.: Macromolecules *15*, 492 (1982)

Editor: K. Dušek
Received May 30, 1985

Author Index Volumes 1–76

Allegra, G. and *Bassi, I. W.:* Isomorphism in Synthetic Macromolecular Systems. Vol. 6, pp. 549–574.
Andrews, E. H.: Molecular Fracture in Polymers. Vol. 27, pp. 1–66.
Anufrieva, E. V. and *Gotlib, Yu. Ya.:* Investigation of Polymers in Solution by Polarized Luminescence. Vol. 40, pp. 1–68.
Apicella, A. and *Nicolais, L.:* Effect of Water on the Properties of Epoxy Matrix and Composite. Vol. 72, pp. 69–78.
Apicella, A., Nicolais, L. and *de Cataldis, C.:* Characterization of the Morphological Fine Structure of Commercial Thermosetting Resins Through Hygrothermal Experiments. Vol. 66, pp. 189–208.
Argon, A. S., Cohen, R. E.. Gebizlioglu, O. S. and *Schwier, C.:* Crazing in Block Copolymers and Blends. Vol. 52/53, pp. 275–334
Arridge, R. C. and *Barham, P. J.:* Polymer Elasticity. Discrete and Continuum Models. Vol. 46, pp. 67–117.
Aseeva, R. M., Zaikov, G. E.: Flammability of Polymeric Materials. Vol. 70, pp. 171–230.
Ayrey, G.: The Use of Isotopes in Polymer Analysis. Vol. 6, pp. 128–148.

Bässler, H.: Photopolymerization of Diacetylenes. Vol. 63, pp. 1–48.
Baldwin, R. L.: Sedimentation of High Polymers. Vol. 1, pp. 451–511.
Balta-Calleja, F. J.: Microhardness Relating to Crystalline Polymers. Vol. 66, pp. 117–148.
Barton, J. M.: The Application of Differential Scanning Calorimetry (DSC) to the Study of Epoxy Resins Curing Reactions. Vol. 72, pp. 111–154.
Basedow, A. M. and *Ebert, K.:* Ultrasonic Degradation of Polymers in Solution. Vol. 22, pp. 83–148.
Batz, H.-G.: Polymeric Drugs. Vol. 23, pp. 25–53.
Bell, J. P. see *Schmidt, R. G.:* Vol. 75, pp. 33–72.
Bekturov, E. A. and *Bimendina, L. A.:* Interpolymer Complexes. Vol. 41, pp. 99–147.
Bergsma, F. and *Kruissink, Ch. A.:* Ion-Exchange Membranes. Vol. 2, pp. 307–362.
Berlin, Al. Al., Volfson, S. A., and *Enikolopian, N. S.:* Kinetics of Polymerization Processes. Vol. 38, pp. 89–140.
Berry, G. C. and *Fox, T. G.:* The Viscosity of Polymers and Their Concentrated Solutions. Vol. 5, pp. 261–357.
Bevington, J. C.: Isotopic Methods in Polymer Chemistry. Vol. 2, pp. 1–17.
Bhuiyan, A. L.: Some Problems Encountered with Degradation Mechanisms of Addition Polymers. Vol. 47, pp. 1–65.
Bird, R. B., Warner, Jr., H. R., and *Evans, D. C.:* Kinetik Theory and Rheology of Dumbbell Suspensions with Brownian Motion. Vol. 8, pp. 1–90.
Biswas, M. and *Maity, C.:* Molecular Sieves as Polymerization Catalysts. Vol. 31, pp. 47–88.
Biswas, M., Packirisamy, S.: Synthetic Ion-Exchange Resins. Vol. 70, pp. 71–118.
Block, H.: The Nature and Application of Electrical Phenomena in Polymers. Vol. 33, pp. 93–167.
Bodor, G.: X-ray Line Shape Analysis. A. Means for the Characterization of Crystalline Polymers. Vol. 67, pp. 165–194.
Böhm, L. L., Chmeliř, M., Löhr, G., Schmitt, B. J. and *Schulz, G. V.:* Zustände und Reaktionen des Carbanions bei der anionischen Polymerisation des Styrols. Vol. 9, pp. 1–45.

Bovey, F. A. and *Tiers, G. V. D.:* The High Resolution Nuclear Magnetic Resonance Spectroscopy of Polymers. Vol. 3, pp. 139–195.

Braun, J.-M. and *Guillet, J. E.:* Study of Polymers by Inverse Gas Chromatography. Vol. 21, pp. 107–145.

Breitenbach, J. W., Olaj, O. F. und *Sommer, F.:* Polymerisationsanregung durch Elektrolyse. Vol. 9, pp. 47–227.

Bresler, S. E. and *Kazbekov, E. N.:* Macroradical Reactivity Studied by Electron Spin Resonance. Vol. 3, pp. 688–711.

Bucknall, C. B.: Fracture and Failure of Multiphase Polymers and Polymer Composites. Vol. 27, pp. 121–148.

Burchard, W.: Static and Dynamic Light Scattering from Branched Polymers and Biopolymers. Vol. 48, pp. 1–124.

Bywater, S.: Polymerization Initiated by Lithium and Its Compounds. Vol. 4, pp. 66–110.

Bywater, S.: Preparation and Properties of Star-branched Polymers. Vol. 30, pp. 89–116.

Candau, S., Bastide, J. and *Delsanti, M.:* Structural. Elastic and Dynamic Properties of Swollen Polymer Networks. Vol. 44, pp. 27–72.

Carrick, W. L.: The Mechanism of Olefin Polymerization by Ziegler-Natta Catalysts. Vol. 12, pp. 65–86.

Casale, A. and *Porter, R. S.:* Mechanical Synthesis of Block and Graft Copolymers. Vol. 17, pp. 1–71.

Cerf, R.: La dynamique des solutions de macromolecules dans un champ de vitesses. Vol. 1, pp. 382–450.

Cesca, S., Priola, A. and *Bruzzone, M.:* Synthesis and Modification of Polymers Containing a System of Conjugated Double Bonds. Vol. 32, pp. 1–67.

Chiellini, E., Solaro R., Galli, G. and *Ledwith, A.:* Pptically Active Synthetic Polymers Containing Pendant Carbazolyl Groups. Vol. 62, pp. 143–170.

Cicchetti, O.: Mechanisms of Oxidative Photodegradation and of UV Stabilization of Polyolefins. Vol. 7, pp. 70–112.

Clark, D. T.: ESCA Applied to Polymers. Vol. 24, pp. 125–188.

Coleman, Jr., L. E. and *Meinhardt, N. A.:* Polymerization Reactions of Vinyl Ketones. Vol. 1, pp. 159–179.

Comper, W. D. and *Preston, B. N.:* Rapid Polymer Transport in Concentrated Solutions. Vol. 55, pp. 105–152.

Corner, T.: Free Radical Polymerization — The Synthesis of Graft Copolymers. Vol. 62, pp. 95–142.

Crescenzi, V.: Some Recent Studies of Polyelectrolyte Solutions. Vol. 5, pp. 358–386.

Crivello, J. V.: Cationic Polymerization — Iodonium and Sulfonium Salt Photoinitiators, Vol. 62, pp. 1–48.

Davydov, B. E. and *Krentsel, B. A.:* Progress in the Chemistry of Polyconjugated Systems. Vol. 25, pp. 1–46.

Dettenmaier, M.: Intrinsic Crazes in Polycarbonate Phenomenology and Molecular Interpretation of a New Phenomenon. Vol. 52/53, pp. 57–104

Dobb, M. G. and *McIntyre, J. E.:* Properties and Applications of Liquid-Crystalline Main-Chain Polymers. Vol. 60/61, pp. 61–98.

Döll, W.: Optical Interference Measurements and Fracture Mechanics Analysis of Crack Tip Craze Zones. Vol. 52/53, pp. 105–168

Doi, Y. see *Keii, T.:* Vol. 73/74, pp. 201–248.

Dole, M.: Calorimetric Studies of States and Transitions in Solid High Polymers. Vol. 2, pp. 221–274.

Donnet, J. B., Vidal, A.: Carbon Black-Surface Properties and Interactions with Elastomers. Vol. 76, pp. 103–128.

Dorn, K., Hupfer, B., and *Ringsdorf, H.:* Polymeric Monolayers and Liposomes as Models for Biomembranes How to Bridge the Gap Between Polymer Science and Membrane Biology? Vol. 64, pp. 1–54.

Dreyfuss, P. and *Dreyfuss, M. P.:* Polytetrahydrofuran. Vol. 4, pp. 528–590.
Drobnik, J. and *Rypáček, F.:* Soluble Synthetic Polymers in Biological Systems. Vol. 57, pp. 1–50.
Dröscher, M.: Solid State Extrusion of Semicrystalline Copolymers. Vol. 47, pp. 120–138.
Drzal, L. T.: The Interphase in Epoxy Composites. Vol. 75, pp. 1–32.
Dušek, K. and *Prins, W.:* Structure and Elasticity of Non-Crystalline Polymer Networks. Vol. 6, pp. 1–102.
Duncan, R. and *Kopeček, J.:* Soluble Synthetic Polymers as Potential Drug Carriers. Vol. 57, pp. 51–101.

Eastham, A. M.: Some Aspects of the Polymerization of Cyclic Ethers. Vol. 2, pp. 18–50.
Ehrlich, P. and *Mortimer, G. A.:* Fundamentals of the Free-Radical Polymerization of Ethylene. Vol. 7, pp. 386–448.
Eisenberg, A.: Ionic Forces in Polymers. Vol. 5, pp. 59–112.
Eiss, N. S. Jr. see Yorkgitis, E. M. Vol. 72, pp. 79–110.
Elias, H.-G., Bareiss, R. und *Watterson, J. G.:* Mittelwerte des Molekulargewichts und anderer Eigenschaften. Vol. 11, pp. 111–204.
Elsner, G., Riekel, Ch. and *Zachmann, H. G.:* Synchrotron Radiation Physics. Vol. 67, pp. 1–58.
Elyashevich, G. K.: Thermodynamics and Kinetics of Orientational Crystallization of Flexible-Chain Polymers. Vol. 43, pp. 207–246.
Enkelmann, V.: Structural Aspects of the Topochemical Polymerization of Diacetylenes. Vol. 63, pp. 91–136.
Entelis, S. G., Evreinov, V. V., Gorshkov, A. V.: Functionally and Molecular Weight Distribution of Telchelic Polymers. Vol. 76, pp. 129–175.
Evreinov, V. V. see Entelis S. G. Vol. 76, pp. 129–175.

Ferruti, P. and *Barbucci, R.:* Linear Amino Polymers: Synthesis, Protonation and Complex Formation. Vol. 58, pp. 55–92.
Finkelmann, H. and *Rehage, G.:* Liquid Crystal Side-Chain Polymers. Vol. 60/61, pp. 99–172.
Fischer, H.: Freie Radikale während der Polymerisation, nachgewiesen und identifiziert durch Elektronenspinresonanz. Vol. 5, pp. 463–530.
Flory, P. J.: Molecular Theory of Liquid Crystals. Vol. 59, pp. 1–36.
Ford, W. T. and *Tomoi, M.:* Polymer-Supported Phase Transfer Catalysts Reaction Mechanisms. Vol. 55, pp. 49–104.
Fradet, A. and *Maréchal, E.:* Kinetics and Mechanisms of Polyesterifications. I. Reactions of Diols with Diacids. Vol. 43, pp. 51–144.
Franz, G.: Polysaccharides in Pharmacy. Vol. 76, pp. 1–30.
Friedrich, K.: Crazes and Shear Bands in Semi-Crystalline Thermoplastics. Vol. 52/53, pp. 225–274.
Fujita, H.: Diffusion in Polymer-Diluent Systems. Vol. 3, pp. 1–47.
Funke, W.: Über die Strukturaufklärung vernetzter Makromoleküle, insbesondere vernetzter Polyesterharze, mit chemischen Methoden. Vol. 4, pp. 157–235.

Gal'braikh, L. S. and *Rigovin, Z. A.:* Chemical Transformation of Cellulose. Vol. 14, pp. 87–130.
Galli, G. see Chiellini, E. Vol. 62, pp. 143–170.
Gallot, B. R. M.: Preparation and Study of Block Copolymers with Ordered Structures, Vol. 29, pp. 85–156.
Gandini, A.: The Behaviour of Furan Derivatives in Polymerization Reactions. Vol. 25, pp. 47–96.
Gandini, A. and *Cheradame, H.:* Cationic Polymerization. Initiation with Alkenyl Monomers. Vol. 34/35, pp. 1–289.
Geckeler, K., Pillai, V. N. R., and *Mutter, M.:* Applications of Soluble Polymeric Supports. Vol. 39, pp. 65–94.
Gerrens, H.: Kinetik der Emulsionspolymerisation. Vol. 1, pp. 234–328.
Ghiggino, K. P., Roberts, A. J. and *Phillips, D.:* Time-Resolved Fluorescence Techniques in Polymer and Biopolymer Studies. Vol. 40, pp. 69–167.
Godovsky, Y. K.: Thermomechanics of Polymers. Vol. 76, pp. 31–102.
Goethals, E. J.: The Formation of Cyclic Oligomers in the Cationic Polymerization of Heterocycles. Vol. 23, pp. 103–130.
Gorshkov, A. V. see Entelis, S. G. Vol. 76, pp. 129–175.

Graessley, W. W.: The Etanglement Concept in Polymer Rheology. Vol. 16, pp. 1–179.
Graessley, W. W.: Entagled Linear, Branched and Network Polymer Systems. Molecular Theories. Vol. 47, pp. 67–117.
Grebowicz, J. see Wunderlich, B. Vol. 60/61. pp. 1–60.
Greschner, G. S.: Phase Distribution Chromatography. Possibilities and Limitations. Vol. 73/74, pp. 1–62.

Hagihara, N., Sonogashira, K. and *Takahashi, S.:* Linear Polymers Containing Transition Metals in the Main Chain. Vol. 41, pp. 149–179.
Hasegawa, M.: Four-Center Photopolymerization in the Crystalline State. Vol. 42, pp. 1–49.
Hay, A. S.: Aromatic Polyethers. Vol. 4, pp. 496–527.
Hayakawa, R. and *Wada, Y.:* Piezoelectricity and Related Properties of Polymer Films. Vol. 11, pp. 1–55.
Heidemann, E. and *Roth, W.:* Synthesis and Investigation of Collagen Model Peptides. Vol. 43, pp. 145–205.
Heitz, W.: Polymeric Reagents. Polymer Design, Scope, and Limitations. Vol. 23, pp. 1–23.
Helfferich, F.: Ionenaustausch. Vol. 1, pp. 329–381.
Hendra, P. J.: Laser-Raman Spectra of Polymers. Vol. 6, pp. 151–169.
Hendrix, J.: Position Sensitive "X-ray Detectors". Vol. 67, pp. 59–98.
Henrici-Olivé, G. und *Olivé, S.:* Kettenübertragung bei der radikalischen Polymerisation. Vol. 2, pp. 496–577.
Henrici-Olivé, G. und *Olivé, S.:* Koordinative Polymerisation an löslichen Übergangsmetall-Katalysatoren. Vol. 6, pp. 421–472.
Henrici-Olivé, G. and *Olivé, S.:* Oligomerization of Ethylene with Soluble Transition-Metal Catalysts. Vol. 15, pp. 1–30.
Henrici-Olivé, G. and *Olivé, S.:* Molecular Interactions and Macroscopic Properties of Polyacrylonitrile and Model Substances. Vol. 32, pp. 123–152.
Henrici-Olivé, G. and *Olivé, S.:* The Chemistry of Carbon Fiber Formation from Polyacrylonitrile. Vol. 51, pp. 1–60.
Hermans, Jr., J., Lohr, D. and *Ferro, D.:* Treatment of the Folding and Unfolding of Protein Molecules in Solution According to a Lattic Model. Vol. 9, pp. 229–283.
Higashimura, T. and *Sawamoto, M.:* Living Polymerization and Selective Dimerization: Two Extremes of the Polymer Synthesis by Cationic Polymerization. Vol. 62, pp. 49–94.
Hoffman, A. S.: Ionizing Radiation and Gas Plasma (or Glow) Discharge Treatments for Preparation of Novel Polymeric Biomaterials. Vol. 57, pp. 141–157.
Holzmüller, W.: Molecular Mobility, Deformation and Relaxation Processes in Polymers. Vol. 26, pp. 1–62.
Hutchison, J. and *Ledwith, A.:* Photoinitiation of Vinyl Polymerization by Aromatic Carbonyl Compounds. Vol. 14, pp. 49–86.

Iizuka, E.: Properties of Liquid Crystals of Polypeptides: with Stress on the Electromagnetic Orientation. Vol. 20, pp. 79–107.
Ikada, Y.: Characterization of Graft Copolymers. Vol. 29, pp. 47–84.
Ikada, Y.: Blood-Compatible Polymers. Vol. 57, pp. 103–140.
Imanishi, Y.: Synthese, Conformation, and Reactions of Cyclic Peptides. Vol. 20, pp. 1–77.
Inagaki, H.: Polymer Separation and Characterization by Thin-Layer Chromatography. Vol. 24, pp. 189–237.
Inoue, S.: Asymmetric Reactions of Synthetic Polypeptides. Vol. 21, pp. 77–106.
Ise, N.: Polymerizations under an Electric Field. Vol. 6, pp. 347–376.
Ise, N.: The Mean Activity Coefficient of Polyelectrolytes in Aqueous Solutions and Its Related Properties. Vol. 7, pp. 536–593.
Isihara, A.: Intramolecular Statistics of a Flexible Chain Molecule. Vol. 7, pp. 449–476.
Isihara, A.: Irreversible Processes in Solutions of Chain Polymers. Vol. 5, pp. 531–567.
Isihara, A. and *Guth, E.:* Theory of Dilute Macromolecular Solutions. Vol. 5, pp. 233–260.
Iwatsuki, S.: Polymerization of Quinodimethane Compounds. Vol. 58, pp. 93–120.

Janeschitz-Kriegl, H.: Flow Birefrigence of Elastico-Viscous Polymer Systems. Vol. 6, pp. 170–318.
Jenkins, R. and *Porter, R. S.:* Upertubed Dimensions of Stereoregular Polymers. Vol. 36, pp. 1–20.
Jenngins, B. R.: Electro-Optic Methods for Characterizing Macromolecules in Dilute Solution. Vol. 22, pp. 61–81.
Johnston, D. S.: Macrozwitterion Polymerization. Vol. 42, pp. 51–106.

Kamachi, M.: Influence of Solvent on Free Radical Polymerization of Vinyl Compounds. Vol. 38, pp. 55–87.
Kaneko, M. and *Yamada, A.:* Solar Energy Conversion by Functional Polymers. Vol. 55, pp. 1–48.
Kawabata, S. and *Kawai, H.:* Strain Energy Density Functions of Rubber Vulcanizates from Biaxial Extension. Vol. 24, pp. 89–124.
Keii, T., Doi, Y.: Synthesis of "Living" Polyolefins with Soluble Ziegler-Natta Catalysts and Application to Block Copolymerization. Vol. 73/74, pp. 201–248.
Kennedy, J. P. and *Chou, T.:* Poly(isobutylene-*co*-β-Pinene): A New Sulfur Vulcanizable, Ozone Resistant Elastomer by Cationic Isomerization Copolymerization. Vol. 21, pp. 1–39.
Kennedy, J. P. and *Delvaux, J. M.:* Synthesis, Characterization and Morphology of Poly(butadiene-g-Styrene). Vol. 38, pp. 141–163.
Kennedy, J. P. and *Gillham, J. K.:* Cationic Polymerization of Olefins with Alkylaluminium Initiators. Vol. 10, pp. 1–33.
Kennedy, J. P. and *Johnston, J. E.:* The Cationic Isomerization Polymerization of 3-Methyl-1-butene and 4-Methyl-1-pentene. Vol. 19, pp. 57–95.
Kennedy, J. P. and *Langer, Jr., A. W.:* Recent Advances in Cationic Polymerization. Vol. 3, pp. 508–580.
Kennedy, J. P. and *Otsu, T.:* Polymerization with Isomerization of Monomer Preceding Propagation. Vol. 7, pp. 369–385.
Kennedy, J. P. and *Rengachary, S.:* Correlation Between Cationic Model and Polymerization Reactions of Olefins. Vol. 14, pp. 1–48.
Kennedy, J. P. and *Trivedi, P. D.:* Cationic Olefin Polymerization Using Alkyl Halide — Alkylaluminium Initiator Systems. I. Reactivity Studies. II. Molecular Weight Studies. Vol. 28, pp. 83–151.
Kennedy, J. P., Chang, V. S. C. and *Guyot, A.:* Carbocationic Synthesis and Characterization of Polyolefins with Si–H and Si–Cl Head Groups. Vol. 43, pp. 1–50.
Khoklov, A. R. and *Grosberg, A. Yu.:* Statistical Theory of Polymeric Lyotropic Liquid Crystals. Vol. 41, pp. 53–97.
Kinloch, A. J.: Mechanics and Mechanisms of Fracture of Thermosetting Epoxy Polymers. Vol. 72, pp. 45–68.
Kissin, Yu. V.: Structures of Copolymers of High Olefins. Vol. 15, pp. 91–155.
Kitagawa, T. and *Miyazawa, T.:* Neutron Scattering and Normal Vibrations of Polymers. Vol. 9, pp. 335–414.
Kitamaru, R. and *Horii, F.:* NMR Approach to the Phase Structure of Linear Polyethylene. Vol. 26, pp. 139–180.
Knappe, W.: Wärmeleitung in Polymeren. Vol. 7, pp. 477–535.
Koenik, J. L. see *Mertzel, E.* Vol. 75, pp. 73–112.
Koenig, J. L.: Fourier Transforms Infrared Spectroscopy of Polymers, Vol. 54, pp. 87–154.
Kolařík, J.: Secondary Relaxations in Glassy Polymers: Hydrophilic Polymethacrylates and Polyacrylates: Vol. 46, pp. 119–161.
Koningsveld, R.: Preparative and Analytical Aspects of Polymer Fractionation. Vol. 7.
Kovacs, A. J.: Transition vitreuse dans les polymers amorphes. Etude phénoménologique. Vol. 3, pp. 394–507.
Krässig, H. A.: Graft Co-Polymerization of Cellulose and Its Derivatives. Vol. 4, pp. 111–156.
Kramer, E. J.: Microscopic and Molecular Fundamentals of Crazing. Vol. 52/53, pp. 1–56
Kraus, G.: Reinforcement of Elastomers by Carbon Black. Vol. 8, pp. 155–237.
Kreutz, W. and *Welte, W.:* A General Theory for the Evaluation of X-Ray Diagrams of Biomembranes and Other Lamellar Systems. Vol. 30, pp. 161–225.
Krimm, S.: Infrared Spectra of High Polymers. Vol. 2, pp. 51–72.

Kuhn, W., Ramel, A., Walters, D. H., Ebner, G. and *Kuhn, H. J.:* The Production of Mechanical Energy from Different Forms of Chemical Energy with Homogeneous and Cross-Striated High Polymer Systems. Vol. 1, pp. 540–592.
Kunitake, T. and *Okahata, Y.:* Catalytic Hydrolysis by Synthetic Polymers. Vol. 20, pp. 159–221.
Kurata, M. and *Stockmayer, W. H.:* Intrinsic Viscosities and Unperturbed Dimensions of Long Chain Molecules. Vol. 3, pp. 196–312.

Ledwith, A. and *Sherrington, D. C.:* Stable Organic Cation Salts: Ion Pair Equilibria and Use in Cationic Polymerization. Vol. 19, pp. 1–56.
Ledwith, A. see Chiellini, E. Vol. 62, pp. 143–170.
Lee, C.-D. S. and *Daly, W. H.:* Mercaptan-Containing Polymers. Vol. 15, pp. 61–90.
Lindberg, J. J. and *Hortling, B.:* Cross Polarization — Magic Angle Spinning NMR Studies of Carbohydrates and Aromatic Polymers. Vol. 66, pp. 1–22.
Lipatov, Y. S.: Relaxation and Viscoelastic Properties of Heterogeneous Polymeric Compositions. Vol. 22, pp. 1–59.
Lipatov, Y. S.: The Iso-Free-Volume State and Glass Transitions in Amorphous Polymers: New Development of the Theory. Vol. 26, pp. 63–104.
Lustoň, J. and *Vašš, F.:* Anionic Copolymerization of Cyclic Ethers with Cyclic Anhydrides. Vol. 56, pp. 91–133.

Madec, J.-P. and *Maréchal, E.:* Kinetics and Mechanisms of Polyesterifications. II. Reactions of Diacids with Diepoxides. Vol. 71, pp. 153–228.
Mano, E. B. and *Coutinho, F. M. B.:* Grafting on Polyamides. Vol. 19, pp. 97–116.
Maréchal, E. see Madec, J.-P. Vol. 71, pp. 153–228.
Mark, J. E.: The Use of Model Polymer Networks to Elucidate Molecular Aspects of Rubberlike Elasticity. Vol. 44, pp. 1–26.
Mark, J. E. see Queslel, J. P. Vol. 71, pp. 229–248.
Maser, F., Bode, K., Pillai, V. N. R. and *Mutter, M.:* Conformational Studies on Model Peptides. Their Contribution to Synthetic, Structural and Functional Innovations on Proteins. Vol. 65, pp. 177–214.
McGrath, J. E. see Yorkgitis, E. M. Vol. 72, pp. 79–110.
McIntyre, J. E. see Dobb, M. G. Vol. 60/61, pp. 61–98.
Meerwall v., E., D.: Self-Diffusion in Polymer Systems. Measured with Field-Gradient Spin Echo NMR Methods, Vol. 54, pp. 1–29.
Mengoli, G.: Feasibility of Polymer Film Coating Through Electroinitiated Polymerization in Aqueous Medium. Vol. 33, pp. 1–31.
Mertzel, E., Koenik, J. LB Application of FT-IR and NMR to Epoxy Resins. Vol. 75, pp. 73–112.
Meyerhoff, G.: Die viscosimetrische Molekulargewichtsbestimmung von Polymeren. Vol. 3, pp. 59–105.
Millich, F.: Rigid Rods and the Characterization of Polyisocyanides. Vol. 19, pp. 117–141.
Möller, M.: Cross Polarization — Magic Angle Sample Spinning NMR Studies. With Respect to the Rotational Isomeric States of Saturated Chain Molecules. Vol. 66, pp. 59–80.
Morawetz, H.: Specific Ion Binding by Polyelectrolytes. Vol. 1, pp. 1–34.
Morgan, R. J.: Structure-Property Relations of Epoxies Used as Composite Matrices. Vol. 72, pp. 1–44.
Morin, B. P., Breusova, I. P. and *Rogovin, Z. A.:* Structural and Chemical Modifications of Cellulose by Graft Copolymerization. Vol. 42, pp. 139–166.
Mulvaney, J. E., Oversberger, C. C. and *Schiller, A. M.:* Anionic Polymerization. Vol. 3, pp. 106–138.

Nakase, Y., Kurijama, I. and *Odajima, A.:* Analysis of the Fine Structure of Poly(Oxymethylene) Prepared by Radiation-Induced Polymerization in the Solid State. Vol. 65, pp. 79–134.
Neuse, E.: Aromatic Polybenzimidazoles. Syntheses, Properties, and Applications. Vol. 47, pp. 1–42.
Nicolais, L. see Apicella, A. Vol. 72, pp. 69–78.
Nuyken, O., Weidner, R.: Graft and Block Copolymers via Polymeric Azo Initiators. Vol. 73/74, pp. 145–200.

Ober, Ch. K., Jin, J.-I. and *Lenz, R. W.:* Liquid Crystal Polymers with Flexible Spacers in the Main Chain. Vol. 59, pp. 103–146.
Okubo, T. and *Ise, N.:* Synthetic Polyelectrolytes as Models of Nucleic Acids and Esterases. Vol. 25, pp. 135–181.
Osaki, K.: Viscoelastic Properties of Dilute Polymer Solutions. Vol. 12, pp. 1–64.
Oster, G. and *Nishijima, Y.:* Fluorescence Methods in Polymer Science. Vol. 3, pp. 313–331.
Otsu, T. see Sato, T. Vol. 71, pp. 41–78.
Overberger, C. G. and *Moore, J. A.:* Ladder Polymers. Vol. 7, pp. 113–150.

Packirisamy, S. see Biswas, M. Vol. 70, pp. 71–118.
Papkov, S. P.: Liquid Crystalline Order in Solutions of Rigid-Chain Polymers. Vol. 59, pp. 75–102.
Patat, F., Killmann, E. und *Schiebener, C.:* Die Absorption von Makromolekülen aus Lösung. Vol. 3, pp. 332–393.
Patterson, G. D.: Photon Correlation Spectroscopy of Bulk Polymers. Vol. 48, pp. 125–159.
Penczek, S., Kubisa, P. and *Matyjaszewski, K.:* Cationic Ring-Opening Polymerization of Heterocyclic Monomers. Vol. 37, pp. 1–149.
Penczek, S., Kubisa, P. and *Matyjaszewski, K.:* Cationic Ring-Opening Polymerization; 2. Synthetic Applications. Vol. 68/69, pp. 1–298.
Peticolas, W. L.: Inelastic Laser Light Scattering from Biological and Synthetic Polymers. Vol. 9, pp. 285–333.
Petropoulos, J. H.: Membranes with Non-Homogeneous Sorption Properties. Vol. 64, pp. 85–134.
Pino, P.: Optically Active Addition Polymers. Vol. 4, pp. 393–456.
Pitha, J.: Physiological Activities of Synthetic Analogs of Polynucleotides. Vol. 50, pp. 1–16.
Platé, N. A. and *Noak, O. V.:* A Theoretical Consideration of the Kinetics and Statistics of Reactions of Functional Groups of Macromolecules. Vol. 31, pp. 133–173.
Platé, N. A. see Shibaev, V. P. Vol. 60/61, pp. 173–252.
Plesch, P. H.: The Propagation Rate-Constants in Cationic Polymerisations. Vol. 8, pp. 137–154.
Porod, G.: Anwendung und Ergebnisse der Röntgenkleinwinkelstreuung in festen Hochpolymeren. Vol. 2, pp. 363–400.
Pospíšil, J.: Transformations of Phenolic Antioxidants and the Role of Their Products in the Long-Term Properties of Polyolefins. Vol. 36, pp. 69–133.
Postelnek, W., Coleman, L. E., and *Lovelace, A. M.:* Fluorine-Containing Polymers. I. Fluorinated Vinyl Polymers with Functional Groups, Condensation Polymers, and Styrene Polymers. Vol. 1, pp. 75–113.

Queslel, J. P. and *Mark, J. E.:* Molecular Interpretation of the Moduli of Elastomeric Polymer Networks of Know Structure. Vol. 65, pp. 135–176.
Queslel, J. P. and *Mark, J. E.:* Swelling Equilibrium Studies of Elastomeric Network Structures. Vol. 71, pp. 229–248.

Rehage, G. see Finkelmann, H. Vol. 60/61, pp. 99–172.
Rempp, P. F. and *Franta, E.:* Macromonomers: Synthesis, Characterization and Applications. Vol. 58, pp. 1–54.
Rempp, P., Herz, J., and *Borchard, W.:* Model Networks. Vol. 26, pp. 107–137.
Richards, R. W.: Small Angle Neutron Scattering from Block Copolymers. Vol. 71, pp. 1–40.
Rigbi, Z.: Reinforcement of Rubber by Carbon Black. Vol. 36, pp. 21–68.
Rogovin, Z. A. and *Gabrielyan, G. A.:* Chemical Modifications of Fibre Forming Polymers and Copolymers of Acrylonitrile. Vol. 25, pp. 97–134.
Roha, M.: Ionic Factors in Steric Control. Vol. 4, pp. 353–392.
Roha, M.: The Chemistry of Coordinate Polymerization of Dienes. Vol. 1, pp. 512–539.
Rostami, S. see Walsh, D. J. Vol. 70, pp. 119–170.
Rozengerk, v. A B Kinetics, Thermodynamics and Mechanism of Reactions of Epoxy Oligomers with Amines. Vol. 75, pp. 113–166.

Safford, G. J. and *Naumann, A. W.:* Low Frequency Motions in Polymers as Measured by Neutron Inelastic Scattering. Vol. 5, pp. 1–27.
Sato, T. and *Otsu, T.:* Formation of Living Propagating Radicals in Microspheres and Their Use in the Synthesis of Block Copolymers. Vol. 71, pp. 41–78.
Sauer, J. A. and *Chen, C. C.:* Crazing and Fatigue Behavior in One and Two Phase Glassy Polymers. Vol. 52/53, pp. 169–224
Sawamoto, M. see Higashimura, T. Vol. 62, pp. 49–94.
Schmidt, R. G., Bell, J. P.: Epoxy Adhesion to Metals. Vol. 75, pp. 33–72.
Schuerch, C.: The Chemical Synthesis and Properties of Polysaccharides of Biomedical Interest. Vol. 10, pp. 173–194.
Schulz, R. C. und *Kaiser, E.:* Synthese und Eigenschaften von optisch aktiven Polymeren. Vol. 4, pp. 236–315.
Seanor, D. A.: Charge Transfer in Polymers. Vol. 4, pp. 317–352.
Semerak, S. N. and *Frank, C. W.:* Photophysics of Excimer Formation in Aryl Vinyl Polymers, Vol. 54, pp. 31–85.
Seidl, J., Malinský, J., Dušek, K. und *Heitz, W.:* Makroporöse Styrol-Divinylbenzol-Copolymere und ihre Verwendung in der Chromatographie und zur Darstellung von Ionenaustauschern. Vol. 5, pp. 113–213.
Semjonow, V.: Schmelzviskositäten hochpolymerer Stoffe. Vol. 5, pp. 387–450.
Semlyen, J. A.: Ring-Chain Equilibria and the Conformations of Polymer Chains. Vol. 21, pp. 41–75.
Sen, A.: The Copolymerization of Carbon Monoxide with Olefins. Vol. 73/74, pp. 125–144.
Sharkey, W. H.: Polymerizations Through the Carbon-Sulphur Double Bond. Vol. 17, pp. 73–103.
Shibaev, V. P. and *Platé, N. A.:* Thermotropic Liquid-Crystalline Polymers with Mesogenic Side Groups. Vol. 60/61, pp. 173–252.
Shimidzu, T.: Cooperative Actions in the Nucleophile-Containing Polymers. Vol. 23, pp. 55–102.
Shutov, F. A.: Foamed Polymers Based on Reactive Oligomers, Vol. 39, pp. 1–64.
Shutov, F. A.: Foamed Polymers. Cellular Structure and Properties. Vol. 51, pp. 155–218.
Shutov, F. A.: Syntactic Polymer Foams. Vol. 73/74, pp. 63–124.
Siesler, H. W.: Rheo-Optical Fourier-Transform Infrared Spectroscopy: Vibrational Spectra and Mechanical Properties of Polymers. Vol. 65, pp. 1–78.
Silvestri, G., Gambino, S., and *Filardo, G.:* Electrochemical Production of Initiators for Polymerization Processes. Vol. 38, pp. 27–54.
Sixl, H.: Spectroscopy of the Intermediate States of the Solid State Polymerization Reaction in Diacetylene Crystals. Vol. 63, pp. 49–90.
Slichter, W. P.: The Study of High Polymers by Nuclear Magnetic Resonance. Vol. 1, pp. 35–74.
Small, P. A.: Long-Chain Branching in Polymers. Vol. 18.
Smets, G.: Block and Graft Copolymers. Vol. 2, pp. 173–220.
Smets, G.: Photochromic Phenomena in the Solid Phase. Vol. 50, pp. 17–44.
Sohma, J. and *Sakaguchi, M.:* ESR Studies on Polymer Radicals Produced by Mechanical Destruction and Their Reactivity. Vol. 20, pp. 109–158.
Solaro, R. see Chiellini, E. Vol. 62, pp. 143–170.
Sotobayashi, H. und *Springer, J.:* Oligomere in verdünnten Lösungen. Vol. 6, pp. 473–548.
Sperati, C. A. and *Starkweather, Jr., H. W.:* Fluorine-Containing Polymers. II. Polytetrafluoroethylene. Vol. 2, pp. 465–495.
Spiess, H. W.: Deuteron NMR – A new Toolfor Studying Chain Mobility and Orientation in Polymers. Vol. 66, pp. 23–58.
Sprung, M. M.: Recent Progress in Silicone Chemistry. I. Hydrolysis of Reactive Silane Intermediates, Vol. 2, pp. 442–464.
Stahl, E. and *Brüderle, V.:* Polymer Analysis by Thermofractography. Vol. 30, pp. 1–88.
Stannett, V. T., Koros, W. J., Paul, D. R., Lonsdale, H. K., and *Baker, R. W.:* Recent Advances in Membrane Science and Technology. Vol. 32, pp. 69–121.
Staverman, A. J.: Properties of Phantom Networks and Real Networks. Vol. 44, pp. 73–102.
Stauffer, D., Coniglio, A. and *Adam, M.:* Gelation and Critical Phenomena. Vol. 44, pp. 103–158.
Stille, J. K.: Diels-Alder Polymerization. Vol. 3, pp. 48–58.
Stolka, M. and *Pai, D.:* Polymers with Photoconductive Properties. Vol. 29, pp. 1–45.
Stuhrmann, H.: Resonance Scattering in Macromolecular Structure Research. Vol. 67, pp. 123–164.
Subramanian, R. V.: Electroinitiated Polymerization on Electrodes. Vol. 33, pp. 35–58.

Sumitomo, H. and *Hashimoto, K.:* Polyamides as Barrier Materials. Vol. 64, pp. 55–84.
Sumitomo, H. and *Okada, M.:* Ring-Opening Polymerization of Bicyclic Acetals, Oxalactone, and Oxalactam. Vol. 28, pp. 47–82.
Szegö, L.: Modified Polyethylene Terephthalate Fibers. Vol. 31, pp. 89–131.
Szwarc, M.: Termination of Anionic Polymerization. Vol. 2, pp. 275–306.
Szwarc, M.: The Kinetics and Mechanism of N-carboxy-α-amino-acid Anhydride (NCA) Polymerization to Poly-amino Acids. Vol. 4, pp. 1–65.
Szwarc, M.: Thermodynamics of Polymerization with Special Emphasis on Living Polymers. Vol. 4, pp. 457–495.
Szwarc, M.: Living Polymers and Mechanisms of Anionic Polymerization. Vol. 49, pp. 1–175.

Takahashi, A. and *Kawaguchi, M.:* The Structure of Macromolecules Adsorbed on Interfaces. Vol. 46, pp. 1–65.
Takemoto, K. and *Inaki, Y.:* Synthetic Nucleic Acid Analogs. Preparation and Interactions. Vol. 41, pp. 1–51.
Tani, H.: Stereospecific Polymerization of Aldehydes and Epoxides. Vol. 11, pp. 57–110.
Tate, B. E.: Polymerization of Itaconic Acid and Derivatives. Vol. 5, pp. 214–232.
Tazuke, S.: Photosensitized Charge Transfer Polymerization. Vol. 6, pp. 321–346.
Teramoto, A. and *Fujita, H.:* Conformation-dependent Properties of Synthetic Polypeptides in the Helix-Coil Transition Region. Vol. 18, pp. 65–149.
Theocaris, P. S.: The Mesophase and its Influence on the Mechanical Behavior of Composites. Vol. 66, pp. 149–188.
Thomas, W. M.: Mechanismus of Acrylonitrile Polymerization. Vol. 2, pp. 401–441.
Tieke, B.: Polymerization of Butadiene and Butadiyne (Diacetylene) Derivatives in Layer Structures. Vol. 71, pp. 79–152.
Tobolsky, A. V. and *DuPré, D. B.:* Macromolecular Relaxation in the Damped Torsional Oscillator and Statistical Segment Models. Vol. 6, pp. 103–127.
Tosi, C. and *Ciampelli, F.:* Applications of Infrared Spectroscopy to Ethylene-Propylene Copolymers. Vol. 12, pp. 87–130.
Tosi, C.: Sequence Distribution in Copolymers: Numerical Tables. Vol. 5, pp. 451–462.
Tran, C. see Yorkgitis, E. M. Vol. 72, pp. 79–110.
Tsuchida, E. and *Nishide, H.:* Polymer-Metal Complexes and Their Catalytic Activity. Vol. 24, pp. 1–87.
Tsuji, K.: ESR Study of Photodegradation of Polymers. Vol. 12, pp. 131–190.
Tsvetkov, V. and *Andreeva, L.:* Flow and Electric Birefringence in Rigid-Chain Polymer Solutions. Vol. 39, pp. 95–207.
Tuzar, Z., Kratochvil, P., and *Bohdanecký, M.:* Dilute Solution Properties of Aliphatic Polyamides. Vol. 30, pp. 117–159.

Uematsu, I. and *Uematsu, Y.:* Polypeptide Liquid Crystals. Vol. 59, pp. 37–74.

Valvassori, A. and *Sartori, G.:* Present Status of the Multicomponent Copolymerization Theory. Vol. 5, pp. 28–58.
Vidal, A. see Donnet, J. B. Vol. 76, pp. 103–128.
Viovy, J. L. and *Monnerie, L.:* Fluorescence Anisotropy Technique Using Synchrotron Radiation as a Powerful Means for Studying the Orientation Correlation Functions of Polymer Chains. Vol. 67, pp. 99–122.
Voigt-Martin, I.: Use of Transmission Electron Microscopy to Obtain Quantitative Information About Polymers. Vol. 67, pp. 195–218.
Voorn, M. J.: Phase Separation in Polymer Solutions. Vol. 1, pp. 192–233.

Walsh, D. J., Rostami, S.: The Miscibility of High Polymers: The Role of Specific Interactions. Vol. 70, pp. 119–170.

Ward, I. M.: Determination of Molecular Orientation by Spectroscopic Techniques. Vol. 66, pp. 81–116.
Ward, I.M.: The Preparation, Structure and Properties of Ultra-High Modulus Flexible Polymers. Vol. 70, pp. 1–70.
Weidner, R. see *Nuyken, O.:* Vol. 73/74, pp. 145–200.
Werber, F. X.: Polymerization of Olefins on Supported Catalysts. Vol. 1, pp. 180–191.
Wichterle, O., Šebenda, J., and *Králíček, J.:* The Anionic Polymerization of Caprolactam. Vol. 2, pp. 578–595.
Wilkes, G. L.: The Measurement of Molecular Orientation in Polymeric Solids. Vol. 8, pp. 91–136.
Wilkes, G. L. see Yorkgitis, E. M. Vol. 72, pp. 79–110.
Williams, G.: Molecular Aspects of Multiple Dielectric Relaxation Processes in Solid Polymers. Vol. 33, pp. 59–92.
Williams, J. G.: Applications of Linear Fracture Mechanics. Vol. 27, pp. 67–120.
Wöhrle, D.: Polymere aus Nitrilen. Vol. 10, pp. 35–107.
Wöhrle, D.: Polymer Square Planar Metal Chelates for Science and Industry. Synthesis, Properties and Applications. Vol. 50, pp. 45–134.
Wolf, B. A.: Zur Thermodynamik der enthalpisch und der entropisch bedingten Entmischung von Polymerlösungen. Vol. 10, pp. 109–171.
Woodward, A. E. and *Sauer, J. A.:* The Dynamic Mechanical Properties of High Polymers at Low Temperatures. Vol. 1, pp. 114–158.
Wunderlich, B.: Crystallization During Polymerization. Vol. 5, pp. 568–619.
Wunderlich, B. and *Baur, H.:* Heat Capacities of Linear High Polymers. Vol. 7, pp. 151–368.
Wunderlich, B. and *Grebowicz, J.:* Thermotropic Mesophases and Mesophase Transitions of Linear, Flexible Macromolecules. Vol. 60/61, pp. 1–60.
Wrasidlo, W.: Thermal Analysis of Polymers. Vol. 13, pp. 1–99.

Yamashita, Y.: Random and Black Copolymers by Ring-Opening Polymerization. Vol. 28, pp. 1–46.
Yamazaki, N.: Electrolytically Initiated Polymerization. Vol. 6, pp. 377–400.
Yamazaki, N. and *Higashi, F.:* New Condensation Polymerizations by Means of Phosphorus Compounds. Vol. 38, pp. 1–25.
Yokoyama, Y. and *Hall, H. K.:* Ring-Opening Polymerization of Atom-Bridged and Bond-Bridged Bicyclic Ethers, Acetals and Orthoesters. Vol. 42, pp. 107–138.
Yorkgitis, E. M., Eiss, N. S. Jr., Tran, C., Wilkes, G. L. and *McGrath, J. E.:* Siloxane-Modified Epoxy Resins. Vol. 72, pp. 79–110.
Yoshida, H. and *Hayashi, K.:* Initiation Process of Radiation-induced Ionic Polymerization as Studied by Electron Spin Resonance. Vol. 6, pp. 401–420.
Young, R. N., Quirk, R. P. and *Fetters, L. J.:* Anionic Polymerizations of Non-Polar Monomers Involving Lithium. Vol. 56, pp. 1–90.
Yuki, H. and *Hatada, K.:* Stereospecific Polymerization of Alpha-Substituted Acrylic Acid Esters. Vol. 31, pp. 1–45.

Zachmann, H. G.: Das Kristallisations- und Schmelzverhalten hochpolymerer Stoffe. Vol. 3, pp. 581–687.
Zaikov, G. E. see Aseeva, R. M. Vol. 70, pp. 171–230.
Zakharov, V. A., Bukatov, G. D., and *Yermakov, Y. I.:* On the Mechanism of Olifin Polymerization by Ziegler-Natta Catalysts. Vol. 51, pp. 61–100.
Zambelli, A. and *Tosi, C.:* Stereochemistry of Propylene Polymerization. Vol. 15, pp. 31–60.
Zucchini, U. and *Cecchin, G.:* Control of Molecular-Weight Distribution in Polyolefins Synthesized with Ziegler-Natta Catalytic Systems. Vol. 51, pp. 101–154.

Subject Index

Acacia gum 18
Adhesive granulating agents 7
Adiabatic deformation 35
— inversion point 55
— thermoelastic measurements 55
Adsorbent activity 168
Adsorption chromatography 131
— energies, differences 164
— isotherm 173
— layer 145
— mode 145
Affine network 51
Agar 24
Agaropectin 24
Agarose 24
Alginic acid 23
— — repeating units 23
— —, tablet binder 24
Amorphous regions 82
— —, intrafibrilar 88
Amyloids 23
Amylopectin 15
Amylose 15
Anisotropy 90, 92
— of linear thermal expansivity 45
Annealing 89
Antitumor agents 27
Arabinorhamnogalactan 29

Bassorin 19
Binary mixture 152, 162
— solvents 153, 156
Bimodal network 67
Bioelastomers 76
Biopolymers 93
Block copolymers 73, 173
— —, microphase-separated 79
Bound rubber 115

Calibration curves 147
Calvet-type microcalorimeter 56
Cancer therapy 29
Carbon black, accessible surface area 118
— —, adhesion index 114, 121

— —, agglomerates 118
— —, aggregates 107, 108
— —, anisometry of particle aggregation 107, 108, 109
— —, A parameter 110
— —, chemisorption 115
— —, DBP number 100, 117
— —, dispersion interactions 120
— —, effective volume 117
— —, free radical 122
— —, grafting 123, 126
— —, graphitization 125
— —, hydrogen loss 124
— —, — transfer 122
— —, particle diameter 107
— —, peripheral hydrogen 122
— —, radial distribution function 108
— —, reinforcement factor R_F 110, 111, 112
— —, secondary structure 108, 119
— —, shape factor 118
— —, solvolysis temperature 115
— —, stiffening effect 107
— —, strength 106
— —, structural parameters 109
— —, structure 107
— —, — concentration equivalence principle 117
— —, surface chemical functions 120, 121
— —, tritium-labeling 123
Carbongel 115
Carboxyl groups 136
Carboxymethylcellulose (CMC) 14
Carrageenan 24
λ-Carrageenan 25
$\varkappa$-Carageenan 25
Cellulose 11
— acetate phthalate 14
—, fibrillar structure 12
—, microcrystalline 11
—, oxidized 13
— powder 11
Chain extensibility, limited 50, 66, 74
— segment 164
Chondrus sp. 25

Chromatographic equation, main 141
Chromatography 147
Cold drawing 78
— —, degree of 84
Composite material 91
Compressibility, anisotropy of 64
Configurations 148
Conformational changes 81
— energy 61, 85
— entropy 43
— gas 49
— properties 170
Constrained junction fluctuation model 51
Contraction, thermal 38
Coordinate, generalized 34
Copolymers, statistical 75
Correction term 59
Correlation parameters 151
Cotton, purified 11
Critical conditions 147, 159, 163, 169, 171, 172
— energy 146, 154
— point 145
— region 146, 155
Cross-linking, degree of 58
— — density 133
— — polymers 134
— — reagent 136
Crystalline polymers, amorphous phase 80
— —, amorphous regions 80, 82
— —, drawn 82
— —, two-phase 91
Crystallization, strain-induced 66, 75
—, stress-induced 44
Cubic lattice 154
Cyclic macromolecules 171
β-Cyclodextrin 10
Cyclodextrin inclusion complexes 10

Deformation 66, 85
—, adiabatic 35
— calorimetry 33, 56, 59
—, residual 74
—, thermodynamics 33
—, type of 58
Dependence, universal 155, 169
Dextran 26
— gels 27
— repeating units 27
Dextrin 16
Disintegrating agent 7
Distribution coefficients 146, 149, 152, 155
— function 91
Domains 75

Echinacea 29
Elasticity, modulus of 89, 90, 92
Elastic force 55

— —, energetic component 50
— —, entropic component 50
— recovery 74
— systems, characteristic ratio 35
— —, ideal energy 35
— —, ideal entropy 36
Elastin 76
Elastomer blends 75
Elastomers, filled 72
—, stretched 82
Elastically active chains 57
Eluent programme 153, 167
Elution power, relative 155, 156, 169
Energy changes 41
— —, intermolecular 86
— contribution 71
— effect, intrachain 57, 58
— inversion, internal 78, 85
Entropy losses 144
Expansivity, anisotropy of linear thermal 45
—, thermal 38
Exclusion, adsorption and critical modes 150
—, critical, adsorption modes 161
— mode 145, 164

Filler 105
Filler-elastomer bonds 124, 125
— - — interactions 114
Filler, morhological characteristics 107
Flory-Huggins lattice 146
Force, generalized 34
Free energy 142
— — change 171
Front-factor 45, 53
FTD, LC of functional oligomers 138ff.
Functional-defective molecules 132
Functional groups 137, 141
— —, irregular 135
— —, regular 135
— —, terminal 148
Functional molecules 163
Functionality 132, 169
—, average 134
—, different 150, 158, 160, 166
— type distribution (FTD) 132ff.
—, weight-average 134
—, zones of different 141
Furcelleran 25
Furcelleria fastigiata 25
Furnace blacks, structural parameters 109

Galactomannans 8, 21
Gas calorimeter 56
Gauche/trans transitions 86
Gaussian networks, thermomechanics 40
Gel point 136
— strength 25

Subject Index

Gels, thermoreversible 21
Gigartina sp. 25
Glassy polymers, oriented 78
Glucans, soluble 28
Gradient and isocratic regimes 137
— elution 166
Graft copolymers 73
Guar gum 22
Gum arabic 18
— —, solubility 19
— —, structure 18
Gums 16
—, sugar components 17
Guth-Gold equation 118

Hard block 70
— phase 73
Heterogeneity, elementary 131
— of polymers 131
Heteropolysaccharides 16
Heteroxylan 29
High-performance liquid chromatography (HPLC) 141
Hydrocolloids, classification 4
—, modified natural 4
Hydroxyethylcellulose 13
Hydroxyl groups 136
Hydroxypropylmethylcellulose 9, 13
Hysteresis 106
— effects 77
—, mechanical 70

Ideal chains 143
Immune response 29
Indian tragacanth 20
Interaction, characteristic radius 144
— energies 144, 169
Interchain effects 58, 61
— interaction 84, 85
Internal energy inversion 85
— standard 157
Intermolecular changes 80
— energy 42
— energy change 41
— interaction 92
Intramolecular energy 42
— — change 41
Inversion 161
— of heat 43
— of internal energy 43
— point, adiabatic 55
—, thermomechanical 44, 46
Isoenergetic chains 54
Isocratic and gradient regimes 137
Isomeric state, rotational 48
— —, theory 60

Junction fluctuations 52

Karaya gum 20
Kinetic parameters 133

Lattice model 170
Linear memory 143
Liquid chromatography 137
— crystalline networks 67
— — polysiloxane 68
Local structure 131
Loading-unloading cycle 70

Macromolecule length 149
— size 131, 156, 161
Macromolecules, cyclic 171
—, non-, mono-, bifunctional 148
—, polyfunctional 171
Membranes 76
Methylcellulose 12
Microfibrils 87
Mobile and stationary phases 141
— phase, interactions 155, 162
— —, polarity 151
Model networks 66
Modulus 106
—, super-high 90
— of elasticity 89, 90, 92
Molecular area 154
— network 39
— sieves 27
— weight 150, 159
— —, equivalent 134
Monolayer volume 168
Mooney-Rivlin equation 64, 119
Mucilages 17

Networks, bimodal 67
—, Gaussian 40
—, liquid crystalline 67
—, model 66
—, molecular 39
—, non-Gaussian 47
—, phantom noninteracting
— reinforced with crystallites 86
—, semicrystalline 81
—, short-chain model 47
—, short-chain unimodal 67
Non-Gaussian effects 75
— — — network 47

Occluded rubber 116
Oligomer chain 170
Oligomers, linear and branched 135
—, polyfunctional 135
Oligobutadiene 136, 147
—, hydroxyl-terminated 163

Ophthalmic products 7
Orientation 91
Overlapping of zones 166

Parameters, kinetic 133
—, physico-mechanical 133
Partition function 143
PDEGA 136, 153, 158
—, linear, cyclic 172
Pectic acid, structure 21
— substances 20
Peterlin-Prevorsek model 89
Phantom noninteracting network 51
Phenomenological equations of state 48
Plantago seed 20
Plant exudates 17
Plastic deformation 85
Polar component 152
Polybutadienes, hydroxyl-terminated 157
Poly(butylene terephthalate) 160
Poly(diethylene glycol adipate) = PDEGA 136, 153, 158
—, linear, cyclic 172
Polyesters, hydroxyl-terminated 157
—, segmented 73
Polymer-filler interaction 69, 72
Polymers, filled 93
Poly(oxypropylene)diol 136
Poly(oxypropylene) triol 136
Polysaccharides 1ff.
Polysaccharides, adhesive granulating agents 7
—, antitumor activity 28
—, — agents 27
— as drug carriers 8
—, bulk-forming laxative thickening agents 22
—, controlled-release tablets 8
—, disintegrating agent 7
—, drug-linked 9
—, emulsions 6
—, enzymatic disintegration 8
— from higher plants 11
—, hetero- 16
—, industrial usage 3
—, liquid dosage forms 6
—, ophthalmic products 9
—, oral dosage forms 6
—, pectic substances 20
—, plant exudates 17
—, properties 5
—, rheological properties 6
—, solid dosage forms 7, 8
—, solubility 5
—, spray inbeddings 22
—, tablets 10
—, topical systems 5
—, viscosity 18
Polysiloxane, liquid crystalline 68

Polystyrene 147
Polyurethanes, segmented 73
Pore 142
— size 149, 156, 161
Pores, slit-like 154
Primitive path model 51
Pseudoplastic properties 26
Psyllium-flea seed 20

Quince seed 20

Random walk 143
Recovery, elastic 74
Reformation 71
Reinforcement 66, 68, 72, 105
Reproducibility 168
Repulsing and attracting walls 143
Restricted volume 143
Retention volume 149, 150, 159, 163
Rotational isomeric state 48
Rubberlike material 40, 68
Rubbers, bound 115
—, filled 71
—, occluded 116
Rupture energy 106

Sarkoma 180 28
SBR 114
Segment size 144
Selectivity 166
Semicrystalline polymers 80
Separating system, model 141
Separation 159, 160, 172
Shrinkage 82
—, thermal 39
Side chains 134
Silicon rubbers 70
Single crystal 79
Sliplink model 51
Slippage 71
Snyder's correlative approach 151
Sodium alginate 23
Solvent power 152
— strength 151
— quality 170
Solvents 152
—, binary 153, 156
Starch 14
—, pharmaceutically used 15
—, properties 14
—, soluble 15
Stationary phases 141
Statistical copolymers 75
— sum 148
Sterculia gum 20
Sterical restrictions 52
Stiffness 106
Stored energy function 49

Subject Index

Strain amplification 119, 120
— -energy function 36
Stress-induced crystallization 44
— softening 69
Surface, interaction with 137
Surfactants 6
Suspending agent 21

Tablet 10
— binder 24
Telechelic polymers 132
Terminal groups 148, 149
— segment 164
Thermal contraction 38
— expansivity 38
— —, negative 38, 82, 83
— shrinkage 39, 83
Thermodynamic potentials 34
Thermodynamics of deformation 33
Thermoelastic measurements, adiabatic 55
— —, isometric 55
— —, isothermal 56
Thermoelastoplastics 68, 70
Thermomechanical equation of state 36
— — — —, statistical theory 40
— inversion 37, 44, 46, 62

Tie-molecules 83
— - —, interfibrilar 87
Tobolsky-Shen semiempirical equation 45, 65
Torsion 46
—, energetic component 46
—, entropic component 46
Tragacanth gum 19
Tragacanthin 19
Transitions, gauche/trans 86
Tube model 51, 53
Two-phase model 83

Unimodal networks, short chain 67
Unperturbed dimensions of chains 60, 61

Vibrational entropy 42
Volume dilation 43, 45, 49, 63
— elasticity 81
Van der Waals equation of state 65

Water content 167
Water-gelatin system 94

Xanthan 26
Xanthomonas campestris 26
Xyloglucans 23